Anonymous

Useful Plants

Plants Adapted for the Food of Man

Anonymous

Useful Plants
Plants Adapted for the Food of Man

ISBN/EAN: 9783744644662

Printed in Europe, USA, Canada, Australia, Japan

Cover: Foto ©berggeist007 / pixelio.de

More available books at **www.hansebooks.com**

USEFUL PLANTS.

PLANTS ADAPTED FOR THE FOOD OF MAN
DESCRIBED AND ILLUSTRATED.

"The eyes of all wait upon Thee; and Thou givest them their meat in due season. Thou openest Thine hand, and satisfiest the desire of every living thing."
—PSALM cxlv. 15, 16.

LONDON:
T. NELSON AND SONS, PATERNOSTER ROW;
EDINBURGH; AND NEW YORK.

1870.

PREFACE

THE object of the series entitled, "OUR WANTS, AND HOW THEY ARE SUPPLIED," of which the present volume is the precursor, is to furnish the young reader with a Library of Useful Information on Common Things, conveyed in a pleasant and agreeable manner. It will have for its subjects the general relations in which man stands to the world around him; his dependence upon its fruits and its animals for the comforts and luxuries of his daily life; the manner in which he has been led to render it available for increased comfort and additional luxury; the inexhaustible resources which it places within his reach; and its obvious adaptation by a Supreme Power to his wants and necessities. It will deal with articles that constantly come under his notice, and yet of which he, as a rule, knows little or nothing; so true it is that familiarity breeds contempt, and that we feel no interest in what is most immediately connected with us. It will endeavour to do this in an easy and attractive style; neither stooping below nor rising above a child's intelli-

PREFACE.

gence; and always endeavouring to convey, though without direct religious teaching, a sense of the debt of love and gratitude due from man to his Creator.

Each volume of the series will be complete in itself, though connected with others by a general scope and aim.

The present is devoted to an exposition of the immense varieties of Vegetable Food which the goodness of God has placed at our disposal. It traces the different climates and countries from which we obtain these varieties, and shows that each country has its appropriate growth, so that man shall never be without food in the world;—so that man, moreover, shall always have cause to exclaim with the Psalmist: "Praise be to Thee, O Father! Thou preparest us food, when Thou hast provided for it!"

CHAPTER I.

Summer in the Harvest-field—About Wheat and Barley—Oats and Oatmeal—Universal Distribution of Corn—All about Rice, its Cultivation, and its Varieties—The Divine Beneficence—Whence comes the Sugarcane?—A Sugar Plantation—Consumption of Sugar.................. 9

CHAPTER II.

The Useful and the Beautiful—Chat about Cherries—Fruits and Juices—Apples and Pears—The Peach—Plums and Damsons—Varieties of the Orange—Ho! for the Strawberry!—The Late-lingering Raspberry—A Dream of the South—Wine and the Vine—The Pomegranate of the Bible—Who likes Pine-apples?..................................... 35

CHAPTER III.

History of the Potato—Precepts and Peas—The Bean and the Onion—Everybody to his Taste—Making up a Salad—Its Ingredients—Cabbages and Superstitions—Tropical Plants—Spices and Condiments—Whence Comes your Dinner?—Exports and Imports—Beverages—Tea, Coffee, and Chocolate 55

CONTENTS.

CHAPTER IV.

The Cocoa-nut Tree and its Manifold Uses—A Story with a Meaning—The Date Palm—Turning Common Things to Use—The Fruit of the Date and its Value—A Coral Island—The Bread-fruit Tree—The Banyan or Indian Fig—The Baobab—Graves for Sorcerers—The Garden of Eden—A Look at Home—Chestnut Trees—An Extraordinary Tree—The Pleasures of Nutting—Walnuts—Brazil Nuts—Almonds—Ingredients of Beer.. 83

CHAPTER V.

Lichens and Mosses—Iceland Moss—Tripe de Roche—Sea-weeds—The Fungi—Mushrooms and Truffles—Geographical Distribution of Plants—Europe, and its Three Regions—Asia—Africa—America—South America: its Llanos, Virgin Forests, and Pampas—New Zealand and Australia—Te Deum Laudamus 126

OUR VEGETABLE FOOD.

CHAPTER I.

BRIGHT is the sky above us; a deep bright blue, with only a few snow-white fleecy clouds speckling its surface to show by contrast how very blue it is. A golden glory seems to rest upon the earth, and tint the leaves of the spreading beech; while those of the oak and the elm and the chestnut are passing from a beautiful purple into many-changing hues of red and brown. The air is bland and warm—so warm, that the cattle quit the open sunshine, and retire into the shade of the thick hedgerows; or stand knee-deep in the waters of the pleasant stream, that flows with a sound of music across the grassy lea. So warm, that the patient horse looks about with wistful eyes for some place of shelter, and endeavours with restless tail to free

THE CATTLE STAND KNEE-DEEP IN THE PLEASANT STREAM.

himself from the cloud of gnats that fret him with their tiny stings. So warm, that the butterfly comes forth to enjoy his brief life of sunshine; and the bee hums merrily from flower to flower, loading himself with sweets. So warm, that the lark and the thrush and the blackbird are full of happiness, and poised in the pure air, or planted on some leafy branch, pour out their thankful hearts to the God of

THE SUN-BROWN REAPERS.

Love in continuous strains of melody. So warm, that the sun-brown reapers in yonder field pause very frequently in their useful toil, and wipe the beaded moisture from their tawny brows. So warm that you and I, dear boys and girls, may well be content to rest ourselves awhile on this flowery bank, and watch with eager glances the pleasant scene before us.

I call it a pleasant scene; and is there one more pleasant in this wide world of ours than a corn-field in the merry autumn-time, when the stalks are bent with the weight of the golden ears? Is there among all the sweet sounds of Nature any sweeter than the sharpening of the reaper's sickle on his ready hone? Does it not tell us of the infinite goodness of Him who, year after year, causes the earth to bring forth her increase, and feeds the mouth of man with wholesome and abundant food.

What should we do, I wonder, without the *cereals?* Frederick knits his brows, you see, at so strange a word, which he thinks has something very un-English about it. And, in fact, it is a Latin word *made* English, and applied to certain grasses which are capable of being converted into human food.

What? Do men live upon grass?

All kinds of corn are grasses, and when their green blades first shoot above the dark rich soil you would hardly detect them from the common

ABOUT WHEAT AND BARLEY.

herbage. All kinds of corn are grasses, and all kinds of corn are called *cereals*.

Look at yonder field, rippling with quick waves of shade and sunshine, like a golden sea. It rejoices in a glorious crop of *wheat;*—wheat, the most delicious and most valuable of *all* the cereal grasses, and which, in most civilized countries, is the chief article used in making bread. In the next field flourishes another useful cereal grass; the sharp spikes bristling about the ears show that it is—*barley*. Now the flour yielded by barley is not equal in fineness and whiteness, nor as a food, to that which we obtain from wheat; but barley is of no little value, because it can be grown in very cold countries, where its sister cereal will never ripen. In the extreme north

WHEAT.   BARLEY.

of Europe, where winter reigns for half the year, and the snow gathers about the hills and valleys in deep masses. and month after month

THE HARVEST.

the streams are fettered in rigid bonds of ice, barley will grow and thrive, repaying the husbandman's labour with a plentiful harvest.

And where the skies are too cold for barley,

and the soil too rich, another cereal grass may be successfully cultivated,—I mean *oats.* It is, as you know, a very graceful plant, its ears drooping from the stalk like fairy-bells, which the lightest wind sets in airy motion. Oats are largely grown in Scotland; and " oatmeal "— for we do not call their produce *flour* — is esteemed a peculiarly nutritious diet for men as well as cattle. No grain is more valued for feeding horses. The Scotch peasant takes porridge at almost every meal.

OATS.

The Russians obtain from oats a pleasant drink, which they call *quass.* The straw is very useful as fodder. The seeds, when they have undergone a certain preparation, are called *groats*, or *grits*, and in this state are employed in making *gruel*. The old name for groats was "grout;" and long years ago grout was much eaten in England, even by the richest and greatest. An old rhyme runs :—

"King Hardicanuto, 'midst Danes and Saxons stout,
Caroused on nut-brown ale, and dined on *grout.*"

Frederick asks me whether wheat, and barley, and oats, grow all over the world.

No, I answer; some countries are unsuited for the growth of these cereal grasses, which will only endure a certain degree of heat or cold. For instance, they will not flourish on the frost-bound shores of the Polar Sea, nor on the banks of the steaming rivers that wind their way through Tropical forests. But consider now the mercy and the wisdom of God :—

In some form or other corn grows everywhere.

From the bleak and barren wastes of Lapland to the sun-scorched plains of Central India, from the great rolling grassy prairies of North America to the muddy swamps of China, from the green valleys of England and the open straths of Scotland to the hills and glens of the West Indian Islands, from the lofty table-lands of the mighty Himalaya to the sunny levels of the shore of the Atlantic, corn, in some form or other, may be successfully cultivated. The Scotch or English emigrant has carried with him across the seas the seeds of wheat and barley and oats; and now in Canada, and Australia, and New Zealand leagues upon leagues of fertile land are enriched with smiling harvests. But in those hot Tropical countries where the sun has a power unknown in our colder lands, and where periods of excessive heat, which scorch and wither every green thing, are followed by weeks of heavy rain, we find another cereal flourishing—

RICE.

MILLET.

Rice, which, for the largest proportion of mankind, forms the principal article of food.

In that grand New World which was made known to Europe by the genius of Columbus, a cereal called *maize*, or Indian corn, blooms over its wide plains and in its pleasant valleys.

In many Tropical countries, as in the north of China and in Hindustan, *millet* is extensively cultivated; while far away in the

MAIZE. RYE.

bleak north of Asia and Europe, in regions too cold for wheat, and on soils too poor and sandy for any other grain, the husbandman sows his fields with *rye*.

Throughout the world, then, we may say, as David said, "Thou," O Lord, "preparest them corn, when Thou hast so provided for it;" Thou preparest corn under the bitter Northern skies, no less than under the scorching, burning, Tropical sun.

Do you remember now the chief cereals, or principal kinds of corn?

Wheat, barley, oats, rice, maize, millet, rye.

The first three of these are so well known to you that I need not attempt to describe them. In reference to the others, I will add a few words.

Rice supplies the main food of nearly one-third of the human race; that is, of fully three hundred millions (300,000,000) of people. It is supposed to have been originally a native of the East Indies, but it is now cultivated in all parts of the globe where heat and moisture are abundant. For it requires something more than heat—namely, a plentiful supply of water. And this is the way in which it is grown:—

The field is covered with long and narrow trenches, each trench being about eighteen inches distant from the other. The bottom is filled with water nearly a foot deep, and in this water the rice seed sown in even rows. After awhile the seeds germinate—that is, take root, and put forth a tiny green stalk—and the water is then drawn off. But in time a quantity of weeds grow up along with the rice; and to kill these weeds the field is again flooded with water, and allowed to remain in that state for fifteen or sixteen days. All the weeds being killed, the water is drawn off, and the rice allowed to bask in the warm sunshine until nearly ripe, when, for the third and last time, it is re-deluged with water.

The cultivation of rice is, therefore, very unwholesome for the labourers, who, when weeding, are compelled to work up to their knees in mud, and consequently suffer much from ague and fever.

Now let me tell you the principal countries where rice is produced, and when we return home, you can look them out in the map of the world. You will find that they are all *Tropical* countries, where there is no winter, but, instead, a season of heavy rains.

India, China, Cochin China, Siam, Cambodia,

Burmah, Japan, Egypt, South Carolina, Georgia, Louisiana, and Florida.

I should add, however, that rice is successfully cultivated in some parts of Europe, as Lombardy in Italy, and Valencia in Spain. Both these places have hot summers, and are well supplied with water.

The finest rice comes from Carolina, but I suppose there are at least one hundred and twenty (120) known varieties. Rice in the husk or shell is termed *paddy*. The Japanese contrive to brew with it a kind of beer, which they call *saki;* the Hindustanese distil from it an intoxicating drink called *arrack;* and the Chinese make several kinds of *rice wine*. In our own country, it is chiefly used for broths and puddings and light cakes; the straw is employed as straw-plait for hats and bonnets.

Let us now fly across the Atlantic on the wings of fancy, and alight in the broad and apparently boundless plains of North America.

Look at those tall culms or stems, throwing off leaves nearly two feet in length and three inches in breadth, with the brownish-yellow ears, generally two or three in number, hanging just below the middle of the stalk—that is the celebrated plant named *maize*, or Indian corn. In the large varieties the ear is often above a foot long, and as thick as my wrist; in the smaller kinds, it will measure four or five inches in length. The meal is much used as groats; but does not make good bread unless mixed with wheat, rye, or barley. When ground and boiled, however, it forms a very nutritious dish called *hominy;* and its porridge, or *mush*, is held in high repute in North America. Its starch is almost equal to arrow-root, and forms the basis of the Oswego Flour and Patent Corn Flour now so popular in England.

Maize will flourish wherever the summer heat is very great; and has been successfully introduced into the south of Europe, as well as into many Asiatic countries.

Of *millet* there are several kinds, for the name is applied to Guinea grass in Africa; to the warra or cheena of Hindustan; and to the durra, which is also very largely cultivated in the Indian peninsula. In Peru grows the *Maize de Guinea*, whose seeds, after being dried by heat, are converted into a white and agreeable flour. Polish millet is cultivated by the cottagers of Poland, and used in the same way as rice. The Africans have several kinds of millet, all useful for food, such as Uzak, Fundi, and Guinea corn. Then there are the Italian millet, three or four feet high; and the German millet, which is but a tiny shrub. Common millet is a native of the East Indies, but thrives famously in all warm countries. Its grain, which is remarkably nourishing, measures only an eighth of an inch in length. It is used as groats, or in flour mixed with wheat-flour. All the millets are flowering grasses, and more nearly resemble rice than any other cereal. One sort, imported into England from the north of China, is the seed of a grass called *Sugar Sorghe,* or *Chinese sugar millet,* on account of its saccharine or sugary properties.

Rye, as I have already told you, is principally cultivated in the north of Europe, and in some of the colder districts of Asia. It makes a dark, almost black bread, which the Russian peasants live upon, but which *you* would think exceedingly unpleasant. It is also employed in distilling the spirituous liquor called " Hollands."

The varieties are numerous, and some are best fitted for sowing in autumn, others for sowing in spring. The word "rye" means rough or hairy, and its ears are not unlike those of barley, but coarser and larger.

Thus, then, we see that corn is universally diffused; that, in some shape or other, it exists all over the world—in the bleak north as in the genial south. "By these striking adaptations of different varieties of grain, containing the same essential ingredients, to different soils and climates, Providence has furnished the indispensable food for the sustenance of the human race throughout the whole habitable globe; and all nations, and tribes, and tongues can rejoice together as one great family with the joy of harvest."

Praise be to thy name, O Lord! For "Thou crownest the year with Thy goodness, and Thy paths drop fatness. The pastures are clothed with flocks; the valleys also are covered over with corn; they shout for joy, they also sing."

How warm the air is this afternoon! The few snow-white clouds that were sprinkled over the azure heaven have disappeared, and the autumn sun shines full upon the earth. Yet what is *this* warmth—extreme as it seems to us —compared with the intense heat of those luxuriant lands where the *sugar-cane* raises its tall yellow stems, and waves its long ribbon-shaped foliage.

Will I tell you about the sugar-cane? Well, as it is a plant of great value and importance, a short account of it cannot fail to interest you.

First, let me tell you where it grows: in the East Indies, in China, in some parts of the south of Europe, in the warm regions

SUGAR-CANES.

of North America, and in the West Indian Islands. It requires a rich soil and a sunny sky.

The plant is perennial; that is, it does not require to be sown every year, but when cut down throws off fresh canes from the root, so that a sugar-plantation lasts without renewal for several years. It has a tall straight stem, from eight to twelve feet high, with many joints, and about one and a half or two inches thick. For two-thirds of its length this stem is filled with a sweet juicy pith, which yields the substance called sugar. The leaves are narrow, and about four feet long; the flowers form a great pointed cluster, quite three feet in length.

Now, if Frederick wanted to make a sugar-plantation, and understood his business, he would choose a favourable spot, on the slope perhaps of a gentle hill, or in some sheltered hollow; and after the ground had been thoroughly ploughed, he would divide it into regular squares, like a gigantic chess-board. Then he would direct his negro servants—there are no slaves now either in North America or in the West Indies, except in Spanish Cuba—to dig across each square a number of parallel rows, about three feet apart, and in each row the young cane shoots or joints will be planted at intervals of about two feet.

At the end of twelve or fifteen days, the joints having taken root, the young stem of this

PLANTING THE SUGAR-CANE.

valuable grass appears above the soil. As soon as the canes are twelve or fourteen inches high,

Sambo and Joe and Quashee set to work hoeing or ploughing up the weeds, which would other-

SUGAR MILL.

wise draw from the soil the nourishment intended for the plant. This labour will have to be repeated in four or five months' time, and meanwhile the soil is kept well turned up, and if necessary, watered.

At the end of a year, generally in the autumn time, the cane is found ready for cutting. This operation must always be performed before it flowers, or the blossoms will consume a great quantity of the sugary pith. The cutting is made

SUGAR MILL—INTERIOR.

a little above the ground, and the stems are afterwards tied in bundles and conveyed to the mill.

The mill consists of three iron rollers, each three to five tons in weight, standing upright and close to one another. By means of machinery they revolve slowly. Passed between these heavy rollers the canes are completely crushed, like a pebble under the wheel of a steam-engine, and the juice, which is of a sweetish taste, and the colour of dirty water, drops into large vessels of iron or copper, exposed to a constant heat. The scum that rises to the top is skimmed off, and the remainder passes from one vessel to another, each hotter than the preceding, until it boils furiously. The purified liquor must now be drawn off into a pan of great size, where it boils more slowly, and begins to concentrate or thicken, forming a sort of thick gruel. It is then conveyed into earthen pots, and allowed to cool and crystallize into sugar. In this state it is called "raw sugar," and imported into Great Britain, where sugar-refining forms an important trade. The principal refineries are situated at London, Bristol, and Greenock.

In the refining process a small quantity of bullock's blood is used, which, rising to the surface of the dissolved sugar, carries with it all

EMBARKATION OF SUGAR.

SUGAR BOILING.

kinds of impurities, so that they can easily be removed by straining them through filters made of a closely woven cotton cloth. The liquor next filtrates slowly through a layer of charcoal, and afterwards being again boiled until it hardens into crystals, is collected in sugar-loaf moulds. It is then ready for use.

Sometimes the melted sugar, instead of being poured into *moulds,* is poured over *strings,* and left to cool slowly. This makes a dainty which most boys and girls are very fond of—sugar-candy. Another dainty, the sight of which sets Nelly's mouth watering, is barley-sugar; that is, refined sugar melted, and pulled out while cooling. *Molasses,* or *treacle,* is also a produce of the sugar-cane, being that part of the juice which, owing to its coarseness, will not crystallize into sugar. The syrup which drains from refined sugar during the various processes I have described, when reboiled constitutes the "golden syrup" which you like so well on your bread and butter.

We are great sugar-eaters in England; but the Americans beat us in this respect. It has been calculated that in Great Britain the sugar annually consumed is equal to 30 lbs. each for every man, woman, and child. In the United States of America it amounts to 40 lbs.; while in France it is only 4 lbs.; and in Russia only a pound and a half.

CHAPTER II.

AS we have now sauntered long enough about this pleasant cornfield, let us wend our way toward home. If we take the gate on the right, Florence, we shall pass through papa's orchard into the garden, and there we can rest awhile in the arbour.

What a profusion of fruit hangs around! Look at the bright scarlet cherries shining like stars among the green leaves; and those delicious apples, with such a fine blush on the side turned towards the sun! Observe, I pray you, that God in his goodness not only gives us what is *necessary* and *useful*, but also the fruit and the flower to please our taste, to gladden our eye, and to enrich the landscape with bloom and fragrance.

The *cherry* is named from the place where it

A BASKET OF FRUIT.

is supposed to have been first cultivated, an Asiatic town, called Cerasus. It was carried from thence to Rome, and spreading over all Italy, was gradually planted in other European countries. It was introduced into England by the Romans. There are now in existence several kinds of cherry, as the hautbois and the black-heart, and every kind as delicious as it is wholesome. Kent is the county where they flourish most; and in the spring it is a charming sight to see the valleys clothed with the delicate snow-white blossoms. The fruit, besides being used for puddings and desserts, yields a juice which is

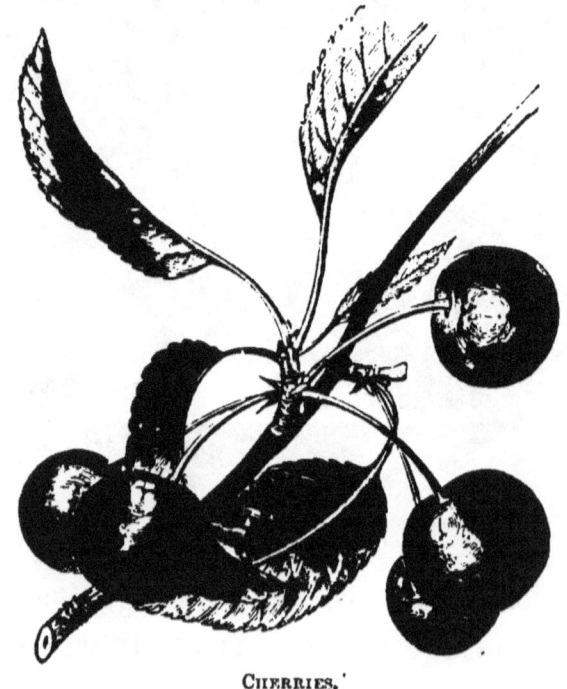

CHERRIES.

employed in making *kirschwasser* (or, cherry-water) and *ratafia*.

I suppose few countries surpass our own in the beauty and abundance of their orchards, and yet it is worth notice that almost all our fruit-trees are of foreign birth. We have trained them, however, with so much care and industry, that they attain to greater excellence in this country than in their native lands.

Apples came originally from Persia; but I do not think a Persian is equal to an English apple; and I am sure the Persians cannot brew any liquor so refreshing and delightful as our Devonshire *cider*, which, I need hardly tell you, is the fermented juice of apples. The fruit is bruised in a stone-trough, and the pulp so obtained collected in hair-cloth bags, and heavily pressed. The juice then runs into a shallow tub, and is stored away in casks to ferment; after which the clear bright cider is carefully strained off.

APPLE.

We get the *pear* from Asia. The best kinds are the *Jargonelle* and the *Bergamot*; but there are more varieties than I should care about repeating. The beverage made from pears is called *perry*.

THE VELVETY PEACH.

Few fruits are superior in delicacy of flavour to the *peach*, and it has a soft velvety skin, of rose-hues and gold, which none can fail to admire. In England it is generally trained on walls; but in the south of Europe is cultivated as a small tree; also in many parts of the East, and in the warmer regions of North and South America. The peach orchards of the United States are of immense extent, frequently containing as many as 20,000 trees. The peach belongs to

PEAR.

PEACH.

the same class or family as the almond, and in leaf and flower the two are much alike. The latter yields a powerful drug called prussic acid, which proves a fatal poison when used in any but the smallest quantities. The *nectarine*, another native of Persia, differs only from the peach in the smooth rind of its fruit. The old Greeks and Romans supposed that their fabled gods drank a wonderfully delicious liquor called *nectar*, and they were so partial to this variety of the peach that they named it the nectar-fruit, or nectarine, as if it were quite as good as the beverage of their deities.

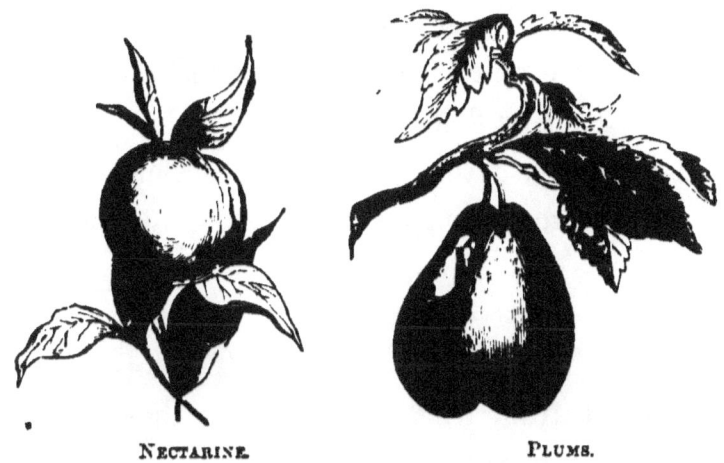

NECTARINE. PLUMS.

Frederick says he likes plums better than peaches. I suspect because he obtains more of the former than of the latter. They are more plentiful, and therefore cheaper. Certainly, the plum is a delicious fruit, and all its varieties— damsons, so called because introduced into England from Damascus, a rich and picturesque city in Syria; green-gages, magnum-bonums, bullaces, egg plums, Orleans plums—are much esteemed for the table. It should be remembered that they are a dangerous fruit when eaten in excess. The Latin name for plum is *prunus*, and dried plums are called *prunes*. These latter are chiefly imported into England from France; which fertile and sunny country, let me add, also sends us dried apples, under the name of Normandy pippins, while Turkey and the Greek Islands send us dried figs.

I wonder whether you and Frederick, or any of your companions, when playing at the old game of

> "Oranges and lemons,
> Say the bells of St. Clements,"

ever think from what far countries those useful fruits are brought? Oranges, lemons, and citrons, all three belong to the same tribe or family; that is, though differing in many respects, they are alike in others, as brother differs from sister, and mother from daughter, and yet a certain resemblance exists between them all.

Oranges were first introduced into Europe from China. The tree is an evergreen, with oblong acute leaves, among which the golden fruit glitters like yellow lamps. Its flowers are white and fragrant, and ancient custom has appropriated them as a fitting ornament for a bride on her marriage-day. It grows in England in the open air, except during winter, when it requires protection, but never attains to such excellence here as in the more genial climes of Spain, Portugal, Malta, and the Azores Islands. There are numerous varieties, as you will not fail to recollect: the bitter Seville orange, which we import from Spain, your mamma uses for marmalade. Its skin supplies a valuable medicine. Red as blood is the pulp of the Malta orange; the Portugal, or Lisbon orange, is known by its thick peel; the Bergamot yields the favourite scent of that name; the St. Michael's is esteemed for its delicious flavour.

ORANGE.

Both the lemon and the citron are natives of the north of India. The former is extensively

THE ORANGE TREE.

cultivated now-a-days in several tropical and subtropical countries on account of its many uses. Its acid juice is valuable in the preparation of lemonade, and is also given as a remedy in various feverish complaints. The calico-printers find it of service in their trade, while the rind, or lemon peel, is an article that no good cook can do without.

The lemon grows in great forests in some parts of Brazil, and it is said the flesh of the cattle which pasture in these woods actually obtains a strong smell of lemons, from their feeding so freely on the fallen fruit.

In our little garden here we have neither citrons nor quinces. The former closely resembles lemons, but are larger, and have a thicker rind. The *quince* is not much grown in England now, but flourishes in the south of Europe, and obtains its name from Cydonia, in the Island of Crete, whence it was first imported. The Portuguese used quinces for making marmalade; and *marmalade* is simply the Portuguese word for a *quince*.

The *strawberry* is a British fruit, which grows wild on our flowery banks and in our verdant hedgerows. The wild variety has a sweet peculiar flavour, but is of very small size. Its name is an old Saxon one, and means the "strawing," or "spreading" berry, because it

QUINCE.

grows on the ground, and creeps along it like ivy over a wall. You do not despise strawberries, I know, Edward. Strawberries and cream, on a hot day, I take to be a dish for a king—or a queen either! Do you like the large kind, called *hautboy*, or *oboe*—that is, *haut bois*, two French words for the "high wood"—because it was first found on the wooded hills of Hungary? Every kind, however, has a rich, full, sugary flavour, and melts in the mouth like snow in the sunshine.

It will soon be time, I see, to prune and tie up the *raspberry* canes; they begin to look

STRAWBERRY.

very ragged. Ah, the glory of the garden is nearly past, and fruit, leaf, and blossom will speedily be numbered with the things that have been! The raspberry lingers long, however, and in the woods you will find the wild plant still bearing fruit. The cultivated raspberry consists of many varieties, and the fruit is either red, yellow, or white. It is not only an agreeable addition to our dessert, but yields a pleasant wine, and in raspberry vinegar a peculiarly grateful and cooling beverage.

RASPBERRY.

But at this rate we shall never get out of the fruit garden. Look at the *gooseberry*-bushes; they are still loaded, though I cannot tell you how many fingers have been picking at them day after day, ever since they began to ripen. I suppose you think the name "gooseberry" has something to do with stupidity; but "goose" is just the same word as "gorse" or "goss," and means "prickly." The plant is a favourite one in England, but not a native; it was imported from Flanders.

Currants are so called, perhaps, because in shape they are not unlike the dried currants, or dried grapes, which are sold by grocers. They have no relationship to the grape, however, but belong to the prickly berry family.

GOOSEBERRIES.

Let us now pass through the vinery into the kitchen-garden. Oh, those beautiful purple grapes! How rich the bloom on every cluster! How graceful the trailing flexible plant, with its broad fantastic leaves and curling tendrils! Does it not carry your mind over the billowy sea to the sweet shining lands of the South, where the hills are covered with vineyards, and you move forward under arches of leaf and fruit, which screen you from the noonday sun, and fill the air with a perfume all their own? Just think, too, of all we owe to the *vine!* First, we have the fresh luscious fruit with its flavour, half sweet, half acid, so grateful to the palate of the invalid. Then, dried in the sun, they are converted into "raisins" and "currants;" the former so called from a Latin word,

CURRANTS.

the latter from Corinth, in Greece, whence we first imported them. The stoneless raisins are known as sultanas, because considered fit for the wife of a sultan; the finest kinds, used for dessert, are so tempting to bees, flies, and other insects, that they have been named *muscatels*, from *musca*, which is just the Latin for "a fly."

GRAPES.

Then, from the grape-juice, when fermented, we procure wine. Our English grapes, however agreeable to the taste, and none are more so, do not make good wine, which, therefore,

VINEYARD.

we are compelled to import from abroad. The different wines are named after the places where they are made.

From wine we procure brandy, which is "burnt wine" flavoured with peach kernels.

Look at the rich crimson blossom of yonder

plant. Is it not beautiful? That is the *pomegranate*, of which you often read in the Bible, and which in old times appears to have been much more valued than is now the case. It flowers, but will not ripen in England; in the south of Europe it flourishes vigorously. Its fruit is as large as an orange, with a thick golden rind just slightly tinged with rose on

POMEGRANATE.

one cheek, like a maiden's blush. The pulp is sweet, delicate, and very cooling. Do you not think it bears a strange name? Well, it simply means "seeded apple." The word "apple" is used for any ball or round-shaped fruit. We speak of the oak-*apple* and the potato-*apple*.

WHO LIKES PINE-APPLES?

PINE-APPLE.

MELON.

In the same way we call the fruit of the ananas, which is a native of Tropical America, *pine-apples*, though they neither grow on pines nor are they shaped like apples. In form, indeed, they resemble the cones of the pine tree. The ananas was introduced into England nearly two hundred years ago, but will not grow in the open air. The Greek word for "apple" is *melon*, whence we derive the name of the fruit called "melons." They are shaped like apples, but much larger. They belong to Eastern countries and to the south of Europe.

CHAPTER III.

COME now to the kitchen-garden, and see how very attractive it looks, with its stock of all kinds of wholesome vegetables, its beds free from stones or weeds, and its paths kept in such admirable order.

Now, Frederick, which vegetable of all you can see do you think the most useful?

The *potato?* I believe you are right. It is almost as necessary to us now as bread. What would you think of a dinner without a good dish of floury potatoes? And yet, three hundred years ago, there was scarcely a potato in all England! This most useful plant, which is a native of Peru, but has long been spread over a great part of America, was brought from Virginia (now one of the United States) by the brave and great Sir Walter Raleigh, in the reign of James I. It had previously been introduced into England by another famous seaman, Sir Francis Drake, but attracted very little attention. It was then called *batata*, and for a hundred years thought only good enough

POTATO PLANT.

for feeding cattle. All at once its cultivation took a sudden start, and soon spread over the entire kingdom, until every cottager in his little garden set aside a corner for potatoes.

It has no *roots*. The straggling stem which

grows underground and bears the potato is really what botanists call a *tuber*.

It is said that the ground which produces 30 pounds of wheat will produce 1000 pounds of potatoes; but the latter are not nearly so nutritious.

Allowing, then, that potatoes are the most useful, which would you call the most *delicious* of vegetables?

Green peas.

Well, I am inclined to agree with you. At all events, it is one of the most graceful.

The pea, a native of the East, was introduced into Europe early in the Middle Ages. Its cultivation now extends over a great part of the known world, and may be successfully undertaken in any climate

GREEN PEA.

not subject to Arctic cold or very early winters.

Peas are removed from the pods and eaten in their fresh green state, or dried and deprived of their husks, and used for pea-soup and pease-pudding. In the latter condition they are sold as *split peas*.

All kinds of grain that grow in pods are called *pulse*. Can you tell me of any other sort of pulse besides peas? *Beans?* Yes; beans are a variety of pulse, and whether *French beans* and *scarlet runners*, which have long narrow pods, or *broad beans*, whose pods are thick and broad, and not fit for eating, they form a useful and agreeable addition to our stores of vegetable food.

In speaking of useful vegetables, we must not omit the *onion;* and while in the neighbourhood of an onion-bed, its strong pungent odour will not allow us to forget it.

It is a bulbous-rooted plant—that is, it has a swelling root, round and smooth, like the bulbs of crocuses, snowdrops, and tulips. The bulb is the part generally eaten, though while the plant is young we also use the emerald-green leaves. Its native country is not known ; but it has been cultivated in Egypt and India for long, long ages, and thence introduced into Europe and America. There are few civilized countries in which the onion is not valued, and its nutritious properties are really very considerable. The Spanish varieties are usually and justly regarded as

ONIONS.

the finest, not only on account of their size, but their more delicate flavour.

How great the variety of vegetables which the all-wise God has created for the behoof and benefit of man! Every taste may be satisfied; every want supplied. We are not limited to a few kinds, but enjoy an almost boundless choice. Our garden here, like most private kitchen-gardens, is laid out on a small scale; yet, see how many different plants flourish within its borders—all more or less distinguished by their agreeable savour and admirably useful qualities.

For instance, besides the common onion, there are *leeks* and *shalotes;* if you do not like *pars-*

nip, you can have *asparagus;* or, instead of either, you may garnish your table with *turnips* and *carrots.* Do you wish for a salad? There

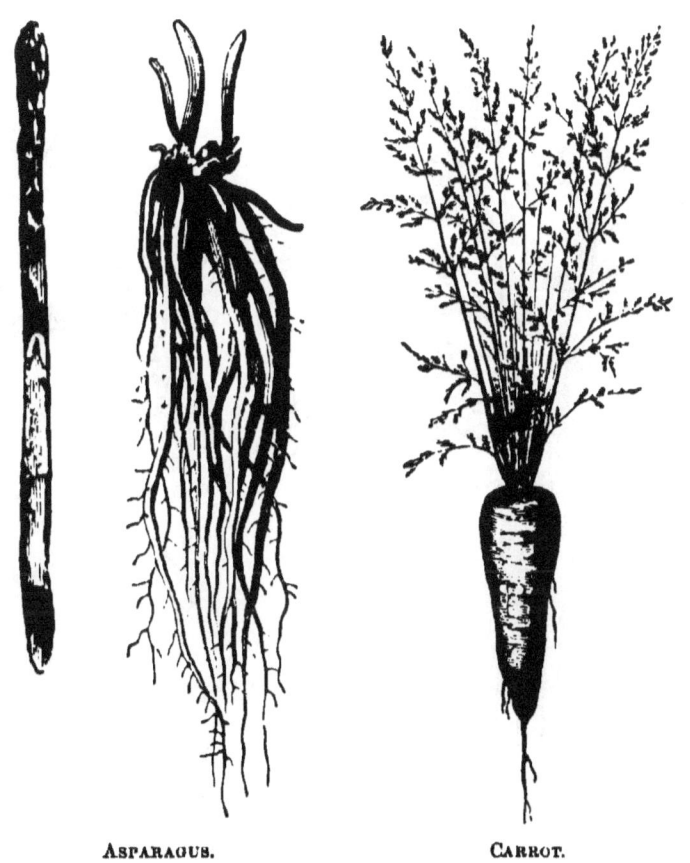

ASPARAGUS.   CARROT.

are *lettuces, mustard* and *cress, endive,* and *celery,* to which you may add some slices of the bright scarlet *beet-root,* though in many parts of Europe sugar is largely extracted from that useful plant. *Radishes* will also lend a zest to your simple meal, and to your bread and butter you will find *watercresses* a delightful accom-

BEETROOT.   RADISH.

paniment. *Cabbages* will figure on your dinner-table—either as greens, summer-cabbages, Brussels sprouts, sea-kail, broccoli, or cauliflower. Both the latter are the flower-beds of two different kinds of "kole," or cabbage. *Spinach* is much relished by many palates; and to others *artichokes* are decidedly acceptable.

SPINACH.

I must tell you something more about cabbages, however; they are so valuable and so abundant a vegetable.

They were well known to the ancients, and very probably it was the Romans who first introduced them into our own country. The Romans called them *brassicæ*, from a Latin word meaning "cut off at the stem;" and also *Caules*, from the quality of the stalk, hence the English word *cole*, or *colewort*. How largely they made use of them we may infer from a verse by an old Latin poet, who, speaking of cabbages, describes them as,—

> "The plants which o'er the world abundance bring
> To peasant humble and to haughty king,—
> Cabbage in winter, and green sprouts in spring."

They cherished some peculiar notions respecting the supposed sympathies of plants, and were of opinion that some liked and some hated their neighbours. An old Roman writer on natural history, called Pliny, says : " The colewort and the vine cherish so deadly a hatred to each other, that if a vine stand near a colewort, you may readily perceive that the former shrinks away from the other ; and yet this wort, which causes the vine thus to retire, if it happen to grow near marjoram, will soon wither and perish in its turn."

What Pliny thought to be an antipathy between the two plants is easily explained. When two plants are placed side by side that require the same juices from the soil for their support and nourishment, the weaker must yield to the stronger—that is, to the one that has the greater power to absorb, or drink up, the valuable moisture.

The varieties of cabbages, or *brassicæ*, now grown in England are very numerous, as :

cauliflower, which is said to have come from the island of Cyprus; *broccoli*, described as an accidental mixture of the common cabbage and the cauliflower; and *sea-kale*, which flourishes,

CAULIFLOWER.

as its name implies, in the neighbourhood of the sea-coast.

Do you not think the greatest epicure might find something to please him among such an abundance? Happily, most kinds of vegetables are within the reach of the poor as well as the rich. Certainly the most valuable are very easily cultivated in any odd corner of ground, if due care be taken, and if the soil be suitable. Potatoes, cabbages, onions, turnips, leeks, lettuces, and the like, will flourish almost everywhere.

How different a garden would look in a very hot country! Instead of cabbages and turnips, you would see rare and curious plants whose names you knew not; and the air, instead of being perfumed with the scent of peas and beans, would smell pungently of spices. Such plants as pepper, turmeric, cloves, and nutmegs, only come to perfection in Tropical lands—that is, in countries where the summer is long, and the heat excessive, because the sun is almost always shining above them.

Pepper, for instance, is the dried fruit of creeping plants, which grow in some islands of the Indian Ocean, in the West Indies, and various parts of South America.

Common pepper, or *black pepper*, which cook uses for flavouring her broths and soups, is a native of the East Indies, and a rambling, climbing plant about twelve feet long, with broad leathery leaves, and an abundance of flowers. The fruit is about the size of a pea, and when ripe, of a bright-red colour. As the plant grows best in the shade, it is therefore trained about a tall pole, under the canopy of a leafy tree. It bears fruit in about three or four years. The berries when gathered are spread on mats, and separated from their stems or spikes by rubbing with the hands or treading with the feet, after which they are exposed to a current of air to clean them. Black pepper consists of the berries thus dried and become black and wrinkled. White pepper is the seed freed from the skin and fleshy part of the fruit by being soaked in water and then rubbed.

PEPPER.

Cayenne pepper is obtained from the *capsicum* plant, which grows in Cayenne, in Guiana. Chillico, which are used in sauces and as a pickle, are also a kind of capsicum.

Where do nutmegs come from?

At one time the *nutmeg tree* grew only in the Banda Islands; but now it is successfully cultivated in several Tropical regions, as at Brazil, Jamaica, Trinidad, Madagascar, and in India. It is about twenty-five feet high, with oblong leaves and scanty flowers, the fruit of the size and appearance of a roundish pear, and, when ripe, of a fine golden yellow colour. If you were to cut open one of the fruit, you would find inside it a large hard kernel, which is the nutmeg, wrapped in a thin flaky substance called *mace*.

CAPSICUM.

NUTMEG.

Did you ever think what a number of countries supply you with your dinner? You thought your dinner was bought and cooked here in England? Yes; but before it could be bought or cooked, some parts of it must have travelled over leagues of land and water.

Just consider for a moment. Your beef is English, I admit; but the turkey at the other end of the table originally came from Asia. Your salt and mustard are English; but your pepper has crossed the seas from Jamaica. Your bread is English; but the yeast which helped to make it came from Germany.

Papa takes a glass of wine with his dessert. Well, it was made at Xeres, in Spain, and is therefore called sherry. He gives you an orange imported from some Atlantic islands called the Azores, or a branch of muscatel raisins, which ripened under the sun of Turkey. The olives are the fruit of a beautiful green tree that flourishes in Spain and the north of Italy; they were brought over to England pickled in salt and water. Your salad oil has been pressed out of Spanish and Italian olives, as you may know by its names, olive oil, Lucca oil, Florence oil. These purple grapes, which look so luscious and tempting, were plucked from a vineyard in the south of France. The

OLIVE TREE AND FRUIT.

butter and the cheese are English produce, but they owe their reddish tint to a colouring stuff called annatto, which is obtained from the seeds of a Tropical plant—a plant that flourishes in the East and West Indies.

Look at those buns. They were made of English flour? True; but they are flavoured with saffron, and saffron is the dried flowers of the *saffron crocus*, which grows in Spain. The word "saffron" is Arabic, and means *yellow;* "crocus," Greek, and means "thread." Saffron looks very much like small pieces of yellow thread.

You are fond, I know, of blanc-mange made with *arrowroot;* but arrowroot is the farina, or flour, of a West Indian plant; and *tapioca*, which cook works up into such insubstantial puddings, is the produce of the milky juice of a South American plant—the *cassava*—which is for the South Americans what the bread-fruit tree is for the Polynesians, yams for the West Indians, or wheat for us English people—the "staff of life."

Now, if you reflect a little, you will see that your dinner has come from a great number of countries: Turkey, Spain, France, Italy, Jamaica, India, South America; and I have mentioned but a few of the dishes generally placed on a gentleman's table.

If you partake of boiled leg of mutton, you

ask for caper sauce. Where will you get your *capers?* They are the buds of a small creeping plant, which is cultivated in several European countries, as well as in Asia and Africa. The finest kind grow in the south of France, and they are pickled before being exported to England.

You know the meaning of the word *exported?* It signifies "sent out of," and *imported,* "brought into."

We speak of the "Exports" and "Imports" of a country. The imports are brought from other lands, and consist of articles which we ourselves do not grow or manufacture; the exports are articles of our own growth or produce, which we dispose of to foreign countries.

I spoke of the *olive* a minute or two ago. It is a tree of singular beauty, frequently forty feet high, and attains to a venerable age. Its leaves resemble those of a willow; are of a dark deep green above and a whitish-gray beneath. The fruit is greenish, whitish, violet, or even black, and about the size of a pigeon's egg. The wood takes a beautiful polish, and being marked and spotted in a very handsome manner, is much used for ornamental work.

The olive tree was greatly valued by the ancient Greeks, and a crown of olive leaves was the reward which they bestowed on any supremely deserving person. An olive branch was also the emblem of peace.

It flourishes abundantly in Palestine and other parts of the East, and there occur numerous allusions to it in the Bible.

We have been talking about Dinner, and how we had to resort to foreign countries if we would set out our table comfortably. But if we could do without their help at dinner-time, we certainly could not dispense with it at tea or breakfast.

We must send to China or India for our *tea*.

What is tea?

A very graceful shrub, which grows spontaneously—that is, without cultivation—in the hilly parts of China, and attains to a height of four or six feet. Its principal product is its leaves, which are of a beautiful dark green on the upper side, and a pale green on the lower.

Having found out its valuable qualities some hundreds of years ago, the Chinese have ever since cultivated it carefully, and many other nations seeing in it an important source of national wealth, it has been introduced into East India, Brazil, the Mauritius, and other genial regions. The European climate is too cold and uncertain to favour its growth.

TEA.

The use of tea as a beverage was introduced into Europe about the middle of the seventeenth century. It was at first regarded with much suspicion, and ridicule was lavishly poured out upon all tea-drinkers, but it gradually made its way until the tea-pot became the ornament of every quiet home, and old age and youth, the rich and the poor, learned to love the cup that cheers but not inebriates.

The grocers sell you numerous varieties —Gunpowder, Hyson, Congo, Pekoe, Souchong, and the like—but they all come from the same plant, and differ only in their mode of preparation, and in the condition of the leaves at the time they are plucked.

The Chinese gather three crops every year: the first, in spring, yields the finest quality; the second takes place a month later; and the third when the leaves are fully grown.

As soon as the leaves have been picked from the plant, they are wetted and dried several times in succession—the drying being effected on hot plates of iron, which causes them to roll up and assume a crumpled form. They are afterwards placed on a table and rolled with the hands. Their colour is then a dull green.

For *black* teas, the leaves are spread out in the air for some time after they are gathered; they are then tossed about till they lose their stiffness; are roasted for a few minutes; are rolled; again exposed to the air for some hours in

TEA FARM.

a moist, soft condition; and, lastly, dried slowly over charcoal fires until the black colour is fairly brought out.

TEA DRYING.

The tea farms, we are told, lie mostly in the north of China, and are generally of small size. They require great attention, as the plant only thrives properly in a rich or well manured soil. The spaces between the shrubs, which are four feet apart, require to be kept in excellent order, and perfectly free from weeds. The soil, too, must be thoroughly drained.

COFFEE AND ITS VARIETIES.

If we obtain our tea from China and India, in the hilly districts where a dry sunny climate prevails, we must go for *coffee* to Tropical regions with a moist atmosphere, such as Arabia, Abyssinia, Brazil, and the West Indies.

The five principal varieties are—

Mocha, a small gray bean, which comes from Arabia; *Java*, or *East Indian*, a large yellow bean, which comes from Hindustan; *Jamaica*, a greenish bean, not so large as the Java, which comes from the West Indies; *Surinam*, a very large bean, imported from the South American coast; and *Bourbon*, a pale yellow bean, brought from the Island of Bourbon, in the Pacific Ocean.

The coffee-tree was originally a native of Arabia and Abyssinia. It is pyramidal in shape, about ten feet high, with branches spread out horizontally almost to the ground. Its leaves are leathery, shining, and evergreen; the flowers small and snow-white, with a delicious perfume. When ripe the fruit is of a deep scarlet colour, and the Arabians call the seeds *bunns*, a word which we have corrupted into *beans*, though they are no more like beans than they are like berries. Before being made use of they are roasted, and ground into powder.

Is there any other breakfast beverage, I

COFFEE.

wonder, for which we are indebted to foreign countries?

Cocoa and *chocolate*, you suppose.

Yes; but both these are the produce of one and the same tree, the *Theobroma cacao*—can you recollect the name?—which grows in the warm districts of South America and the West Indies. It is but a little tree, seldom growing

COFFEE PLANTATION.

taller than an apple or pear-tree, but it divides into many branches. The fruit is from six to eight inches long, yellow, except on the side next the sun, which wears a rosy tint, and is shaped like a cucumber. The rind is thick and warty, the pulp sweetish and not unpleasant. In this pulp lie numerous almond-like seeds, with a thin, pale, and reddish-brown shell, enclosing a bitter, dark-brown, oily, aromatic kernel.

CACAO POD.

We call things *aromatic* when they give out a pleasant pungent smell, like nutmeg, cinnamon, carraway, or other spices.

The seeds I speak of are the *cocoa beans* sold in our shops; when bruised and broken up into small pieces, they are called *cocoa nibs*. To make *chocolate*, they are reduced to a fine paste, and mixed with spices and pounded sugar. The word is Mexican (*chocolatl*, from *choco*, cocoa, and *lotl*, water). Chocolate was first introduced into Europe about 1520.

The cacao-tree attains its full growth in seven or eight years, and generally yields two crops annually. The fruit, when gathered, is allowed to ferment for five days, either in heaps on the ground, or in earthen vessels. It is then opened by the hand, and the seeds dried by the sun or by fire.

Chocolate, I may tell you, is not only a very agreeable but a very wholesome beverage. It contains a great deal of nutriment—that is, of feeding matter; and for breakfast or tea, is now in great request. I wish it was cheaper, for the poor would find it a valuable addition to their scanty stock of comforts.

CHAPTER IV.

THE cacao you must not confuse with the *cocoa-nut tree*, for not the slightest point of resemblance exists between them, except that both flourish best in warm countries. The latter is one of the numerous species of palms which abound in the Tropical parts of the world, and which are all known by their tall straight trunk, and crown of broad long leaves.

The cocoa-nut tree is generally of the thickness of a man's body, and from sixty to one hundred feet high. The leaves droop with a curve, and measure from twelve to twenty feet in length. The nuts grow in clusters of five to fifteen, and as many as eighty or a hundred nuts will be found on one tree.

I suppose the cocoa-nut tree is the most useful in the world, and it is really astonishing how valuable is every part of it. Nothing of it is thrown away.

Take, first, the *fruit*:—

The *fibre*, or hairy-like substance, of its husk, or outer shell, makes excellent cordage, brooms, and brushes.

The *shell* is fashioned into ladles, cups, and goblets.

Cocoa Nut.

The *kernel* affords a nutritious food; the oil pressed out of it is used in manufacturing candles, or as a lamp oil, or to anoint the body; and the milk is a refreshing beverage.

The *sap*, when fermented, gives a liquor called patna wine, and a spirit called arrack. The *juice* is frequently boiled down to yield sugar.

The *dried leaves* are used for thatch, and woven into screens, mats, baskets, and sails.

When young, the middle part of the *stem* is sweet and eatable; the *topmost bud*, or palm cabbage, is relished as a great delicacy.

COCOA-NUT TREE.

The *wood* is very hard, and takes a beautiful polish.

I do not wonder that you open your eyes so wide, Frederick. It never occurred to you that any one tree could serve such numerous purposes. And in this we may see the goodness and wisdom of God; for where the cocoa-nut thrives best, few other trees are found, and the inhabitants of those distant islands which were its native birth-place must consequently have suffered much distress but for the manifold usefulness of this famous tree.

A very clever writer has put this fact before us in a striking and impressive manner. He says,—

" Imagine a traveller passing through one of those countries, situated beneath a glowing sky, where shade and coolness are seldom to be enjoyed; and habitations, in which to obtain the rest so needful and so welcome to the traveller, are only to be met with at long intervals. Weary and sick at heart, the poor wayfarer at length descries a modest hut, surrounded by trees with straight and lofty stems, each stem crowned by a gigantic tuft or crest of great leaves—some upright and others drooping—the whole giving a peculiar character and air of beauty to the scene.

"At the spectacle, the traveller's spirits revive; summoning up all his energies, he hurries towards the cabin, and is soon beneath its hospitable roof. He expects to obtain nothing but repose, and a shelter from the burning sun; but he soon finds himself mistaken. His host makes him welcome, provides him with a seat, and perceiving that he is athirst, offers him a sourish drink: it refreshes his parched tongue like nectar.

"After he has rested himself awhile, the Indian invites him to share his meal. Various meats are served up before him in a brown-looking vessel, smooth and glossy; there is also a supply of wholesome wine of a particularly agreeable flavour. Towards the close of the repast he is provided with a delicious dessert, and is made to taste some excellent spirits. Wondering and amazed, the traveller addresses his host: 'Whence,' he says, 'whence, in this desolate country, do you obtain such a variety of provisions and liquors?'

"'I have but one resource,' replies the Indian.

"'And what may that be, for it seems inexhaustible?'

"'My cocoa-nut tree,' is the answer.

"'What! a cocoa-nut tree? Wine, spirits, sweets, meat—can you procure *all* from a single tree?'

"'Yes: the water which I offered to you on your arrival I draw from the fruit before it ripens, and some of the nuts which contain it weigh three or four pounds each.

"'The almond, whose flavour you so much admired, is the mature fruit.

"'The milk, whose taste was so pleasant to your palate, I also draw from the nut; this cabbage, which you found so delicate, was the top of the palm tree—but I rarely indulge in so great a treat, because the tree from which the cabbage is cut dies shortly afterwards.

"'The wine in which you did me the honour to drink my health was also furnished by the cocoa-nut tree. To procure it, I make an incision into the *spathe*, or envelope, of the flowers, and immediately a white liquor issues,

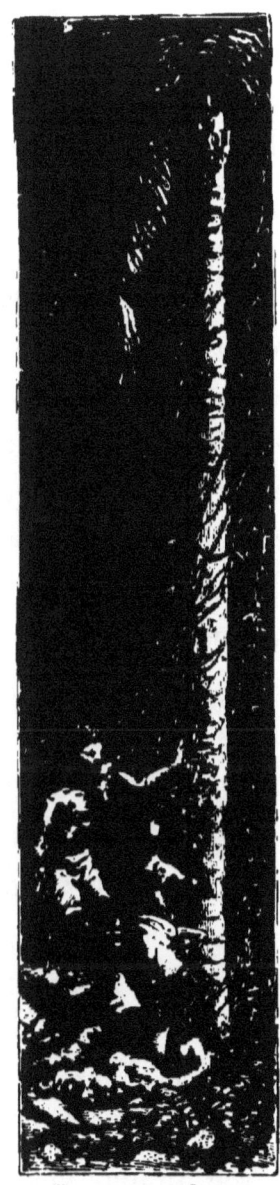

TAPPING THE PALM.

which is gathered into proper vessels, and called *palm wine*. Exposed to the sun it turns sour, and is converted into vinegar.

"'By a process of distilling, I make the capital brandy with which you finished off your repast. The sap supplied the sugar which sweetened our preserves.

"'The dishes, bowls, and utensils we have made use of were all fashioned from the shell of the nut.'

"'What a wonderful tree!' exclaims the traveller.

"'But I have not yet done. This hut, under whose roof we are sheltered, I owe entirely to this useful palm. With its wood I have constructed its sides; with its leaves, dried and plaited, have formed the roof; these leaves, woven into an umbrella, protect me from the sun in my walks; and their filaments, properly treated, provide me with clothes. These mats, useful for so many purposes, are also manufactured from them. The sifter which you behold was ready made to my hand in that part of the tree whence the leaves issue; woven together, the leaves may be used as sails for ships; the kind of fibre enveloping the fruit is excellent caulking for their seams, and makes excellent thread, and all kinds of cord and cables. Finally, the oil that seasoned some of our dishes, and that which now burns in my lamp, is pressed out of the fresh kernel.'"

Another kind of palm, found in the dry sandy deserts of Arabia and Egypt, is scarcely less remarkable in this respect. I mean the *date palm*.

I shall give you a description of it from an old writer and traveller, Pococke. You know what its fruits are like, for you have eaten them frequently.

After being planted, it is three or four years before much of it appears above the ground.

A DELICATE FOOD.

If the top is cut off, with the boughs springing from it, the young bud and the ends of the

DATE PALM.

tender boughs form a delicate food, something like chestnuts, but much finer, and sold very dear.

The boughs are of a grain like cane; and when the tree grows larger, a great number of stringy fibres seem to stretch out from the boughs on each side, crossing each other in such a manner as to form a sort of bark like close net-work; and this the natives spin out with the hand, and with it make cords of all sizes, which are mostly used in Egypt. They also make of it a sort of brush.

Of the leaves they make mattresses, baskets, and brooms; and of the branches all kinds of cage-work, as well as square baskets for packing, that serve for many uses instead of boxes. The ends of the boughs that grow next to the trunk, being beaten and pounded like flax, the fibres separate, and, when tied together at the narrow end, serve excellently for brooms.

These boughs do not fall off of themselves in many years, even after they are dead, as they die after five or six years; but, as they are of great use, the natives commonly cut them every year, leaving the ends of them on the tree, which strengthens it very much. When, after many years, they drop off, the whole tree is weakened by their loss, and frequently overthrown by the wind.

The palm tree grows very high, in one straight stem, and has this peculiarity,—that the heart is the softest and least durable portion, the outer parts being the most solid; so that when the Arab builds his house, he generally uses the trees entire, or divides them only into two parts.

The fruit of the date, when fresh, eats well roasted, and also prepared as a sweetmeat; it

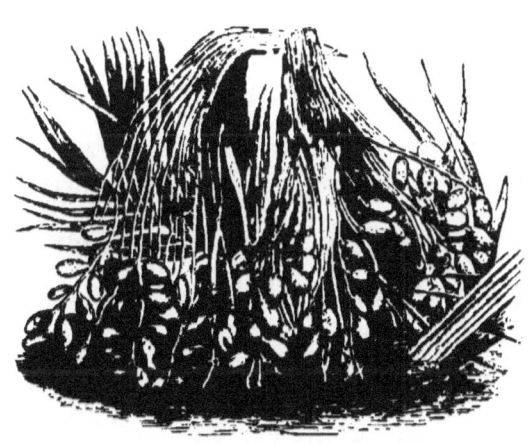

CLUSTER OF DATES.

is esteemed of a hot nature, and as it comes in during the winter, being ripe in November, Providence seems to have designed it as a warm food, during the cold season, to comfort the stomach.

Thus far our old writer and traveller, Pococke. I may add that the date tree, to the Arabs, is even more important than wheat to us, and if it were at any time to fail, millions of our fellow-creatures would perish of famine. In Egypt, in Arabia, in Persia, in Barbary, and the countries bordering on the Great African Desert, dates form the principal food, and date palms the principal wealth of the people. Cakes of dates, pounded and kneaded together until they become so solid as only to be cut with a hatchet, are the provision which the traveller takes with him on his long journey through the barren and rocky Sahara.

Such is the date palm. The traveller, floating slowly along the great Egyptian river, gazes with admiring eyes on its slender shapely trunk and emerald-green crown, as tree after tree is lit up by the beams of the rising or setting sun; and he thinks, perhaps, of the days when Moses meditated in their grateful shadow, or refreshed himself with their pleasant fruit. It is, however, the peculiar tree of the *desert*, and the hearts both of master and of slave are glad when they see from afar, standing out clearly and distinctly against the hot cloudless sky, the tall groves of palm that tell of the presence of some silent pool or of the flow of noisy waters.

It is pleasant—is it not?—while sitting in this cool leafy arbour, under an English sky, to carry our thoughts across the wide seas and the distant lands to the wondrous region of the great stony desert, or the rich glowing forests of the Tropics. What is it enables us to do so? Memory and imagination! And surely these powers of the mind, which thus conquer time and space, are among the greatest and most marvellous gifts of God to man! Memory and imagination! Yes, these are the wings which bear us up while we traverse the valleys and the mountains, the broad plains and the rolling rivers, to gaze upon ever fresh scenes of beauty, majesty, and awe!

Let us, then, take another flight. Frederick, do you feel strong enough?

What do you say? "It is all nonsense, uncle."

No nonsense, but absolute fact. Just now I fancied myself in a wide sandy wilderness, with a burning sun above, and not a drop of water near to quench my thirst or cool my parched lips. And looking across the glaring sand, I thought I saw the forms of some tall trees sharply defined against the sky, and I knew they must be palm trees. Then my soul rejoiced, for wherever the palm tree grows, water will be found.

Well, now I will picture to myself a little island in the great South Sea. All around it runs a ring of coral, broken here and there by little channels of deep blue water, through one of which I paddle in a light canoe, and reaching the shore, contrive to effect a landing. Let us look around. The scene is charming, but very strange. There are hills and valleys and plains, as in our own dear land, yet they look very different to ours, because covered with trees and shrubs, with plants and grasses, such as we have never seen before. What gorgeous blossoms! What broad rich leaves! What tall magnificent ferns! How soft and green the turf on which we tread, and how delicious the fragrance that fills the air! There are deep shady woods before us, and on either hand spread miles of luxuriant pasture, but none of those fields of wheat and bearded barley which enrich the landscapes of England and Scotland. But if you look more closely you will see that wheat and barley are not needed in this fairy island. Yonder trees you will surely recognize: yes, it is the cocoa-nut, which supplies the islanders with an inexhaustible store of wholesome food.

> In many a leafy southern isle,
>   Girdled with sunny seas,
> Which, like a happy infant, smile,
>   When stirred by summer's breeze,
>
> In pitiless wastes of desert sand,
>   Which scorch the genial seed—
> The palms in blest abundance stand
>   To succour human need!
>
> Where'er adventurous man has trod,
>   His Maker goes before;
> We find the mercy of our God
>   Fold round us more and more!

In the islands of the Pacific Ocean and Indian Archipelago grows another tree remarkable for its usefulness. I have already alluded to it—the *Bread-fruit tree*—but I think you will be glad to learn something more about it. Its fruit supplies the chief food of the natives of those regions; its inner bark, nearly all their

BREAD-FRUIT TREE.

clothing; its timber, their huts, furniture, and canoes; its milky juice, their daily beverage.

It is a very slender tree, but grows to a stature of fifty feet, often without throwing off a single branch for half that height. Its leaves are long, glossy, and dark green. The fruit is roundish, about the size of a child's head, and covered with a roughish rind, marked in small square or lozenge-shaped divisions. When fully ripe, its colour is a rich yellow. It hangs to the branches by a short thick stalk, generally in clusters of two or three together.

The fruit is gathered before it is quite ripe, when the pulp is white, mealy, and doughy, like new bread. It is cut into three or four pieces, and the core taken out; a hole is dug in the earth, and some red-hot stones placed at the bottom of it. These are covered with a layer of green leaves, upon which lies a layer of the fruit; then another of hot stones; and leaves and fruit alternately, until the hole is nearly filled; when leaves and earth, some inches thick, are spread over the whole, and the fruit allowed to remain untouched for about half an hour.

When roasted sufficiently the outside appears nicely browned—the inside, white and pulpy, like the crumbs of a new loaf. It has little taste, but is exceedingly nutritious.

Shall we come back again to this pleasant arbour? Yet no; let us take one more flight in Tropical climes. Do you see yonder immense tree, or rather mass of trees, which seem like a labyrinth of shady alleys, or the avenues in an old ancestral park? That is a fig-tree.

Frederick *says* nothing, but his *looks* mean much. They mean: " Were I not a respectful and polite young gentleman, I should say, *I don't believe you*, uncle. Is there not a fig-tree trained against yonder garden-wall? Have I not often noticed its branches, clothed with short hairs, and its long deeply-marked leaves, and its purple pear-shaped fruit? And yet you speak of a fig-tree resembling a number of shady alleys! Pooh! The fig is more like a shrub than a tree."

True, Master Frederick, and yet there flourishes in India another and a far grander species of fig-tree, called the *banyan*, one of which will cover a space sufficient to contain seven thousand (7000) persons! It bears a rich scarlet fruit, like a couple of ripe cherries, and it yields so many other valuable products, and is so

BANYAN.

grand and gorgeous in itself, that the Hindus regard it with especial admiration.

The reason it extends over such a vast surface is simply this : its branches grow downwards, take root in the ground, and in their turn become small trees; so that one trunk will sometimes be surrounded by as many as three hundred and fifty stems (350), until it resembles a veteran captain in the midst of his soldiers. Both the parent tree and its offspring live for ages,—

> " It is a goodly sight to see
>   That venerable tree,
> Far o'er the lawn irregularly spread,
> Fifty straight columns prop its lofty head ;
>   And many a long depending shoot
>   Seeking to strike its root,
> Straight, like a plummet, grows towards the ground.
>
> " Nor weeds, nor briars, deform the natural floor ;
> And through the leafy copse which bowers it o'er
>   Come gleams of chequered light.
> So like a temple doth it seem, that there
> A pious heart's *first* impulse should be *prayer*."

The *Baobab* is called by botanists the *Adansonia digitata*, because it was discovered on the Senegal, in West Africa, by a French botanist named Michel Adanson, and because its leaves are *digitate*, that is, divided like fingers.

It is the largest known tree in the world. I

do not mean that it is the loftiest—for in *height* it is surpassed by many—but that it possesses the most colossal trunk, often measuring from 90 to 100 feet in girth or circumference. Its branches are as thick as our English forest-trees, and frequently 70 feet long, while they form a hemispherical head of 120 to 150 feet in diameter,* like a dome or canopy of foliage. Its flowers are white and exceedingly large; they are attached to drooping stems or flower-stalks about three feet long.

The trunk of the baobab is rugged, and rent into wide furrows, which supply securely-sheltered retreats for sheep and other animals. The branches, as may be supposed from their immense length, extend over a vast area of ground; and the fruit are suspended from their nether surface, clothed with a very rich, deep sea-green, downy substance, which induced Captain Clapperton, the traveller, to compare them to " so many velvet purses." The popular English name for them is " Monkey Bread;" because they are much relished by

* Do you know what is meant by the word *diameter?* Take a stick, and run it through an orange; the stick will represent the diameter of the orange, so far as it is contained within the fruit.

the monkeys, who hide themselves among the thick foliage, and gambol merrily from bough to bough.

The African savages make daily use of the dried leaves of the baobab. They mix them with their food, for the purpose of reducing their excessive perspiration, and counteracting the ardour of their burning climate.

The fruit of the baobab, says the author of the "Vegetable World," is eatable; its flesh is sweet, and of a pleasant flavour; the juice, when extracted and mixed with sugar, forms a beverage very useful in the putrid and pestilential fevers of the country. The fruit is transported into the eastern and southern parts of Africa, and the Arabs convey it to the countries adjoining Morocco, whence it finds its way into Egypt.* The negroes take a portion of the damaged fruit and the woody bark, and burn them for the sake of the ashes, from which, with the help of palm oil, they manufacture soap. A still more curious use is made of the trunk of the baobab; they expose upon it, or

---

* Our young readers are requested to look out on their maps all the places mentioned in these pages.

rather, within it, the bodies of their bards, and of others for whom they consider interment in the earth unfitting.

They pick out a baobab whose trunk has been already attacked and hollowed out by insects or fungi; they enlarge the cavity into a kind of chamber, and therein they suspend the dead body. This done, they close up the entrance to this natural tomb with a plank. In the interior the body becomes perfectly dry, and is, in fact, converted into a mummy without further preparation.

This peculiar kind of burial is especially reserved for a class called the *Guerrots*, who are the musicians and poets of the country, and in the courts of the negro kings preside at all dances and festivals. While they live and flourish, they derive a peculiar influence from this species of talent, and the ordinary negroes look upon them as potent sorcerers, and treat them with distinguished honour. After death, they still regard them with a superstitious reverence; and they fancy that if they consigned the body of one of these sorcerers to the earth, like the bodies of other men, they would

A WOODLAND SHADE.

expose themselves to an everlasting malediction.

Hence, the colossal baobab is chosen as the last resting-place of the Guerrot. There is a strange poetry, says Figuier, in this custom of a barbarous people, which induces them to bury their poets between heaven and earth in the heart of the giant of trees.

Thinking of all the trees and plants that a merciful God has thus placed at the service of man, for purposes of utility or refreshment—of trees and plants differing according to the soil and climate in which they flourish, but all, or nearly all, contributing to human enjoyment—I am reminded of Milton's exquisite description of Paradise, of the garden of Eden, which, so far as trees and plants are concerned, seems still to exist upon earth:—

> "Over head up grew
> Insuperable height of loftiest shade,
> Cedar, and pine, and fir, and branching palm,
> A sylvan scene; and as the ranks ascend,
> Shade above shade, a woody theatre
> Of stateliest view. Yet higher than their tops
> The verdurous wall of Paradise up-sprung;
> Which to our general sire gave prospect large
> Into his nether empire neighbouring round;

> And higher than that wall a circling row
> Of goodliest trees, laden with fairest fruit,
> Blossoms and fruit at once, of golden hue,
> Appeared with gay enamelled colours mixed."

I have said nothing about the *manioc*, the *banana*, or *plantain*—a very delicious saccharine fruit; the *tare-root* of Polynesia, or the useful and widely-spread yam; in a word, I have not half exhausted the vegetable food which an All-wise God has provided for the inhabitants of Tropical countries; but it is time we finished our afternoon "walk and talk." We must come back from those rich, glowing, brilliant lands to our own colder and more sober country, where, however, the mercy of the Creator has been equally bountiful to His creatures. I am sure that I would not give up yonder golden wheat-field for all the cocoa-nut groves in the islands of the Pacific. Besides, the industry and enterprise of our merchants and our seamen bring all the treasures of East and South, and West and North to our doors; and by the skilful contrivances of science, we are enabled to grow, in places provided with an artificial climate, the rarest fruits and plants of every region.

THE BANANA.

If we go forth, however, into our own woods, we shall find several species of vegetable food; not so important as those we have just mentioned, but still of some utility, and assuredly of agreeable flavour. In winter time what would you think of a dessert without a plate of *chestnuts?* And, for that matter, in summer time what would you think of an English wood without its avenue of leafy *chestnut trees?* How glorious a sight they present towards the end of May, when their large clusters of cone-shaped blossoms shine tenderly among their glossy foliage! In Bushey Park, on the Thames, there is a long walk—upwards of a mile, I think, in length—which is planted on either side with three or four rows of vigorous and venerable chestnut trees, and I cannot describe to you the beauty of the scene when they are all in flower; it seems like a landscape borrowed from the fairy world.

The chestnut tree frequently attains the height of ninety to a hundred feet, and a circumference proportionate to its height. It grows rapidly, and lives to a grand old age. Its leaves are large, oblong, smooth, and shining, and deeply indented at the edges; its flowers are composed of several divisions, forming a kind of cone, of a pinkish-white, and enclosed in what is called an involucrum, or envelope, which, in the month of September or October,

Chestnut.

becomes thick and leathery, and assumes a prickly outside. Open it, and within you will discover two, three, or even five curiously shaped fruits, called chestnuts.

In the centre of France and in the Alpine

valleys, these fruits form the principal food of the poor peasantry; and when roasted, they are exceedingly wholesome and nutritious.

Why are they called chestnuts?

That is right, Frederick; never lose a grain of knowledge for want of asking. The common

CHESTNUT TREE OF THE HUNDRED HORSES.

name is said to be of Turkish origin, and to be derived from the Turkish custom of grinding up the nuts and mixing them with the food of horses suffering from disease of the chest.

I have told you that this beautiful tree

sometimes attains the height of a hundred feet, and a circumference in proportion. I suppose the largest in the world—at all events, one of the largest—grows on the green slope of Mount Etna, the great volcanic mountain in Sicily, where it is known as the "Chestnut of a Hundred Horses," because, it is said, a hundred horsemen would find shelter under its branches.

Its girth at three feet above the ground is about a hundred and eighty-four feet. It is entirely hollow now, and supported chiefly by its bark; but the glory of its leafage still endures, and is renewed with every spring. The Sicilians tell a story to the effect that Queen Joanna of Naples, while on a visit to Sicily, ascended Mount Etna, accompanied by all the nobles of the island on horseback. A storm coming on, the queen and her attendants took refuge under this tree, whose vast canopy effectually protected them from the rain.

Mr. Bartlett is of opinion that this extraordinary tree consists, in reality, of several trunks; but other observers are of a different opinion, and maintain that, huge as is its girth, it is only a single chestnut. And it should be added, that many chestnuts on Mount Etna are eighty to ninety feet in circumference.

The chestnut tree belongs to what botanists call a *family* of trees, each species in the order possessing certain characters in common. This order or family is called *Amentaceæ*, and includes many kinds of willow, oak, birch, plane, and poplar. None of these contribute to man's vegetable food; but one of the members of the Amentaceæ is more liberal, as every schoolboy knows—I refer to the hazel.

What boy has not tasted the delights of nutting? Can you imagine any more delightful sport than to sally forth on an autumn evening, armed with a hooked stick to bring the branches within your reach, and while the sunset is flooding earth with its glory of gold and crimson, emerald and purple, to plunge into wood or coppice, where every leaf seems a treasury of beautiful colour, and plunder the hazel of its sweet delicious fruit?

The hazel is not a tree, however, but a shrub, and seldom grows above twelve feet high. Its branches are slender, upright, and flexible; it bears both male and female flowers, arranged in catkins, or cone-shaped clusters, which turn green towards the end of autumn, and flourish until about Christmas. The nut is partly enclosed in a kind of fleshy envelope or involucrum; and while the flavour of the kernel is sweet and tender, the form of the nut is, to my eye, very graceful and attractive.

HAZEL NUTS.

Long years ago, when the common people — and not only the common people, but learned men, and kings and nobles—were very superstitious, it was usual to cut a hazel branch for use as a divining-rod; that is, it was supposed to possess the power of pointing out the secret source of a water-spring, or even the hiding-place of a concealed treasure, by a peculiar vibratory motion, when held in the hand of a properly qualified person.

Let me now direct your attention to another tree, whose fruit is invariably a welcome addition to the dessert. In fact, *walnuts* and wine always go together; though, for my part, I am content with the walnuts *without* the wine, and so, I suppose, would most boys be.

The walnut is a large tree, with a whitish bark, more or less rent and fissured according to its age. Its stem is cylindrical—that is, shaped like a cylinder—and springs upwards to a considerable height before it throws out any branches; these branches are very ample and spreading, so that the tree, with its leafy canopy, wears a very handsome appearance. Its leaves are of a dull green colour. It is not a native of England, though it flourishes in our climate with much vigour, but of Persia, India, and the region of the Caucasus. Its leaves are smooth, but somewhat leathery in texture; like the hazel and the chestnut its flowers are disposed in catkins; and its fruit, as you know,

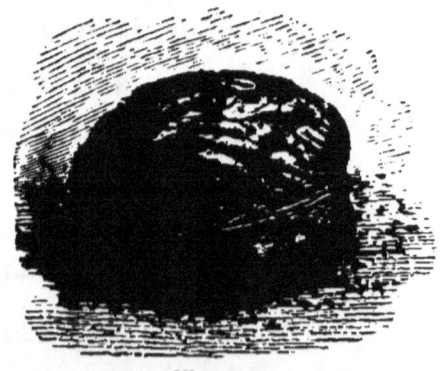
WALNUT.

consists of a hard woody shell, enclosed in a green fleshy husk. Walnut-picking is a fine source of amusement for the young; the fruit communicating to the hands a dark peculiar stain, which converts a fair and ruddy Saxon into a brown Gipsy with extraordinary speed.

Walnuts are brought to the table for dessert when ripe; but they are also used as a delicious pickle, and for this purpose are gathered in an earlier stage.

The walnut tree was known to the Greeks and Romans, and by them, as by us, was valued for its beautiful timber as well as for its agreeable fruit. I cannot tell you when it was first introduced into our own country, but the old botanist Gerard speaks of "its green and tender nuts," when "boyled in sugar," as "a most pleasant and delectable meat, comforting to the stomach, and expelling poison." The tree is largely cultivated in France, Spain, Italy, and Germany, and its nuts enter largely into the food

WALNUT TREE.

of the rural population. Finally, we may sing its praises in the words of Cowley :—

> "Its timber is for various uses good;
> The carver it supplies with useful wood:
> It makes the painter's fading colours last;
> A table it affords us, and repast:
> E'en while we feast, its oil our lamp supplies;
> The rankest poison by its virtue dies."

Where do we get the *Brazil nuts?*

They are the fruit of a colossal tree, called the *Bertholletia excelsa*, which flourishes in the dense and magnificent woods of the Brazil, on the banks of the great river Amazons. We are told that the Indians make their way into the woodland recesses, and adopt every device they can think of to irritate the monkeys, which live in thousands among the trees. In revenge, the monkeys pelt their persecutors with these nuts, which the Indians immediately proceed to collect with all possible speed.

Another favourite nut is the *almond*. This is the fruit of a tree which closely resembles the peach. It is a native of Arabia, but is now cultivated over the whole of Europe; and in our English gardens is valued for the beauty of its delicate, but frail and fleeting blossoms.

THE GRACEFUL HOP.

> "Fair pledges of a fruitful tree,
> Why do you fall so fast?"

But stay: what pretty creeper is this, that mixes its bright-green leaves and dark-brown

ALMOND.

clusters with the honeysuckle bloom, and almost threatens to overpower it?

When talking of man's vegetable food, it would have been unpardonable for me to omit the graceful *hop*.

I need not remind you that beer, which has

been called "the Englishman's beverage," is made with hops and malt, the former giving it a bitter flavour. Malt, as you know, is prepared from barley.

Hop.

The cultivation of the hop was introduced into England in the reign of Henry VIII. The plant is a native of Europe and of some parts of Asia; but as it only thrives in a peculiar soil, good hops are rarely met with. In England it flourishes bravely in Kent, Sussex, Hampshire, and Worcestershire; but those of Kent and Sussex are the best, and not alone the best in England, but the best in the world.

The leaf will remind you of that of the vine; it is either three-lobed or five-lobed—that is, marked out in three or five divisions like the lobes of the human ear. Its flowers grow in loosely spread clusters or cones, resembling the fir-cones in shape, and these gradually enlarge and ripen into the fruit. It is the ripe fruit that is used in brewing.

As the hop will not grow without support, we train it upon tall stout poles, eighteen or twenty feet high; and a hop-garden, filled with regular rows of these, presents a very picturesque scene in autumn time. Hop-picking begins about the middle of September. Women and children are chiefly employed in the pleasant task, but men are required to lower the poles and cut away the plants, which they then deposit across a sort of wooden trough, and the women and children pick off the fruit.

Afterwards, the hops are dried in the *vast*, or kiln, and packed into bags, or *pockets*, ready

HOP-PICKING.

for sale. The *vines*—that is, the stalks—are used for manure in England; but the Swedes manufacture their fibres into a coarse kind of strong white cloth.

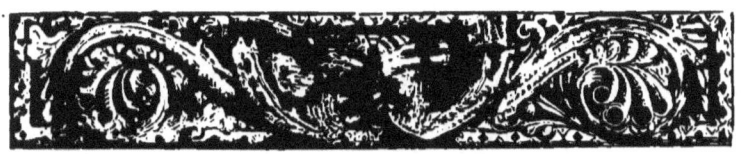

CHAPTER V.

IT is worth remarking from how many and what various sources Man has learned to supply his physical wants, and in what manner he has learned to conquer the hard conditions of Nature. When he cannot avail himself of the beauty of the corn-field, of the golden treasures of the orchard, the wealth which loads the plains and valleys of Tropical regions, he knows how to be content with poorer fare: with the scant berries that ripen in northern climates; with the mosses that clothe the sterile waste, or the sea-weeds cast up on the wave-worn shore.

For example: he is frequently indebted for subsistence to those simple forms of vegetation called *lichens*. Among these, the most valuable, perhaps, is the Iceland Moss (*Cetraria*

ICELAND MOSS.

islandica), which is sold in every chemist's shop. It is the principal article of food which the natives of frozen Iceland possess. In fact, says a popular writer, without it they would as certainly perish as the favoured inhabitants of Britain without the more highly organized cereal plants (as wheat and barley) which, year after year, wave in all their golden beauty over the whole land, and are so strikingly suggestive of Nature's bounty and munificence.

BEARDED LICHENS.

What barley, rye, and oats are to the races of Asia and Western Europe; the olive, the grape, and the fig to the inhabitants of the Mediterranean districts; the date-palm to the Egyptian and Arabian; rice to the Hindu;

and the tea-plant to the Chinese,—the Iceland moss is to the Laplanders, Icelanders, and Esquimaux.

The Iceland moss is found in Scotland as well as in Iceland; but while in the former country it grows but scantily on the summits and sides of the lofty mountains, in Iceland it spreads profusely over the entire surface of the country. On the black lava which forms the soil of the western districts, and in the dreary region which lies at the base of the great volcano of the Skaptar-Jokul, it attains a large size; and, in the summer months, numerous parties wander to these places with all their household effects in order to collect it, either for exportation to the Danish merchants, or for their own use as an article of common food. These excursions, we are told, are usually made once every three years, for the lichen takes three years to grow ripe, after the spots where it flourishes have been cleared.

Olassen and Povelsen—can you read these names?—two travellers in Iceland, inform us, that a person will collect four tons in a week; and they add, that with these four tons of nutritious Iceland moss he is better off than with one ton of wheat.

This seems doubtful; and another authority states, that the meal obtained from it, when mixed with wheat-flour, certainly produces a greater quantity of bread, though, perhaps, of a less nutritious or feeding quality than could be manufactured from the latter alone. I don't think my young readers would like this Iceland moss bread, however, on account of its bitter taste. But the Icelanders and Laplanders manage to get rid of the bitterness, to a great extent, by the following process :—

First, they chop it to pieces; then they pound it for several days in water, mixed with lime or salt of tartar; next, they pour off the water, dry the moss, mix it with the flour of the common knot-grass, make it into a cake, or boil it, and washing it down with reindeer's milk, pronounce it a dish fit for a king!

The powder is very like common starch in appearance, and not unlike it in some of its properties; for it swells in boiling water, and in cooling becomes a fine jelly, which soon hardens into a tough, transparent substance, very pleasant to the taste, and particularly so, as I daresay you will recollect, when flavoured with sugar, milk, a little wine, or spices.

Another lichen, which sometimes enters into the food of man, is called the *Tripe de Roche*— that is, Rock Tripe; and in the cold, bleak, barren regions about the Pole, where winter reigns for nine or ten months of the year, and the sea is covered with ice, and the ground with frozen snow, this lichen is not to be despised. It is bitter and unpleasant to the taste; but for a few days, in the absence of everything else, it helps to support life.

When Franklin and Richardson were exploring the North Polar regions, and endeavouring to make their way across the icy wilderness, from the Coppermine River to a place called Fort Enterprise, their supplies of provisions became exhausted, and they were compelled to turn to the *Tripe de Roche* to satisfy, in some degree, the cravings of hunger. I will quote from the account of their dismal journey a painful passage descriptive of their sufferings:—

"Mr. Hood," says the writer, "was now

nearly exhausted, and obliged to walk at a gentle pace in the rear, Dr. Richardson kindly keeping beside him; whilst Franklin led the foremost men, that he might make them halt occasionally till the stragglers came up. Credit, however, one of their most active hunters, became lamentably weak, from the effects of *Tripe de Roche* upon his constitution; and Valiant, from the same cause, was daily growing thinner. They only advanced six miles during the day, and at night satisfied the cravings of hunger by a small quantity of *Tripe de Roche*, mixed up with some scraps of roasted leather. Having boiled and eaten the remains of their old shoes, and every shred of leather which could be picked up, they set forward at nine, like living skeletons, advancing by inches, as it were, over bleak hills, separated by equally barren valleys, which contained not the slightest trace of vegetation except this eternal *Tripe de Roche*."

The lichens bearing this curious name are black in colour, and leather-like in substance, with small points about them like coiled wire buttons. They grow on the rocks, in cold and exposed situations, and are attached to them by short strong hairs or fibres. Some of them are not unlike shagreen, or shark's skin; while others seem corroded, like a fragment of burnt skin, as if the cliff where they grew had suffered from the action of fire.

GATHERING MOSS IN THE POLAR REGIONS.

"Though they contain," says Mr. Macmillan, "a considerable quantity of starch, they are exceedingly bitter and astringent, and produce intolerable griping pains when eaten. No one would have recourse to them for food except in a case of dire necessity. The Canadian hunters, who are often reduced to the last extremity during their long and toilsome excursions in search of furs, through the desolate regions of Arctic America, often allay the pangs of hunger with this nauseous diet. And sometimes in my own wanderings among the almost unknown and unvisited solitudes of the Scottish mountains, when my stock of provisions was exhausted, and a renewal was not to be expected, the nearest shepherd's sheiling being perhaps many miles distant, I have been compelled to satisfy my cravings by eating small portions of the *Tripe de Roche*, which I found blackening the dreary rocks around. In such situations, I have felt deeply how weak and helpless is man, when thrown forth from the social scenes and comforts of civilized life, left to his own unaided resources, and exposed to the merciless energies of physical nature; and how, without some ultimate trust in the Almighty Source of his being, that being is but a straw upon a whirlpool."

Food for man is scattered everywhere. We

find it on the sea-shore no less than on the Polar rocks. In many parts of Scotland, the fishermen make use of the sea-weed called *dulse*, in the absence of other provision, and it is said to be agreeable in flavour and nutritious in quality. The peasant living inland need

MUSHROOMS.

only go down to the meadow-bank, and gather the excellent fungi that flourish there, to furnish himself and family with a dish that might tempt an epicure.

You ask me, what are *fungi?*

Well, they are flowerless plants, that usually ripen, as it were, in autumn; that do not spring from the *earth*, but grow like parasites on organic bodies, and are nourished by animal and vegetable substances in a corrupted state. You will see them fattening upon the body of the oak, or covering the surface of vinegar and cheese with what is popularly called *mould*, or spreading all over the wall of a damp cellar, for they seem to shun the light, and to prefer moist, close, and badly-ventilated places. Toadstools are fungi, and so are mushrooms; but whilst the former are poisonous, the latter are eminently wholesome. They, however, are not the only kind of fungi which can be used by man as food. The truffle is much esteemed by the lover of dainties, and figures on the dinner-tables of the wealthy. .

The true value of fungi as Vegetable Food is not known in Great Britain, and the mushroom and truffle are the two varieties chiefly sought after. But it is much to be regretted that so abundant and wholesome a provision is allowed to perish year after year in the fields and meadows, when it would supply the poor man's kitchen with a truly delicious bill of fare. Dr. Badham, who has written a capital book on Eatable Fungi, says: "I have myself witnessed whole hundredweights of rich, wholesome food rotting under trees; woods tottering with food, and not one hand to gather it; and this, perhaps, in the midst of potato blight, poverty, and all manner of privations, and public prayers against imminent famine. I have indeed been grieved to see pounds innumerable of extem-

MUSHROOMS AND TOADSTOOLS.

pore beefsteaks (that is, fungi tasting like beefsteaks, and requiring no cooking) growing on our oaks in the shape of *Fistulina hepatica;* *Agaricus fusipes*, to pickle, in clusters under them; puff-balls, which some of our friends have not inaptly compared to sweet-bread, for the rich delicacy of their unassisted flavour; *Hydna*, as good as oysters, which they somewhat resemble in taste; *Agaricus deliciosus*, reminding us of tender lamb-kidneys; the beautiful yellow *Chantarelle*, growing by the bushel, and no basket but our own to pick them up; the sweet nutty-flavoured *Boletus*, in vain calling himself *edulis* (that is, eatable), where there are none to believe him; the dainty *Oredea*, the *Agaricus heterophyllus*, which tastes like the crawfish when grilled; the *Agaricus ruber* and *Agaricus virescens* (Agaricus green and red), to cook in any way, and equally good in all;—these are among the most conspicuous of the edible funguses."

What hard names are given to these plants! I fear my young readers will be unable to pronounce them. But it does not matter about the *names*, so long as the *things* are good; and though an *Agaricus* may be hard to pronounce, it is very easy to eat.

But, remember, it is difficult for an unskilful person to distinguish between fungi that are wholesome, and fungi that are poisonous. Do not attempt to gather them, therefore, without the assistance of some experienced guide. People have lost their lives before now by mistaking the common toadstool for the mushroom, which it is very like; that is, very like in appearance, but not in properties, for while the mushroom is wholesome and delicious, the toadstool is disagreeable and poisonous.

Before I conclude, let me tell you something about the Truffle. Where do you think it grows? Why, *underground.* It is generally found in beechwoods, and grows in clusters six, eight, or twelve inches below the surface of the soil.

You see it can flourish, unlike other plants, without the aid of light.

TRUFFLES.

In appearance it is like a rough potato; at first is white, afterwards black; and when ripe, it is cracked all over and covered with projections, like a pine-cone or a fir-apple.

Internally it is solid, grained like a nutmeg, and of a dirty white or pale brown colour.

In Great Britain, our truffles seldom exceed

three or four ounces in weight; but in many parts of Italy, Germany, and France, they weigh six, eight, and even ten pounds.

You ask me, How are they eaten?

Oh, they are either placed on the table fresh, like fruit, or roasted like potatoes; or dried and cut into slices, to flavour our soups.

But if they grow underground, how do we find them out?

Well, dogs, which, as you know, possess a peculiar power of scent, are trained to hunt for them. Pigs, too, are remarkably fond of truffles, and by the instinct with which God has gifted them contrive to detect the places where they are to be found, and set to work to dig them up. Then in upon the scene comes man, and seizes the spoil before the pigs can carry it off or devour it. It is, however, possible to cultivate them, just as vegetables are cultivated.

Before I conclude this sketch of the various articles which make up man's vegetable food, I think it desirable I should show you in what manner those articles are distributed over the surface of the globe, that you may understand the extent of our obligations to the development of commercial enterprise. But for the labours of our travellers and navigators, our tables here in Britain would be but scantily furnished, for our climate is so changeable, that some of the most delicious fruits and useful vegetables will not ripen here, or can only be brought to maturity through expensive artificial means; and, consequently, had we no adventurous seamen, they would be utterly unknown to the great majority of our population.

Linnæus, the great Swedish botanist, has summed up in a few words the main features of what is called the geographical distribution of plants. He says: "The dynasty of the palms reigns in the warm region of the globe; the tropical zones are inhabited by whole races of trees and shrubs; a rich crown of plants adorns the plains of southern Europe; hosts of Gramineæ (or grasses) occupy Holland and Denmark; numerous tribes of mosses are encamped in Sweden; while the brownish-coloured algæ and the white and gray lichens alone survive in bleak and frozen Lapland—the most remote inhabited spot of earth."

The great principle of this distribution may be expressed in three words : *Like to like.* Only those plants flourish together which possess some important characteristics in common, are fed by the same kind of soil, and encouraged by the same kind of climate. The beech will not grow side by side with the cocoa-nut palm; the breadfruit tree would perish where the cherry is hale and hearty. Plants are most numerous, are richest in foliage, and, as a rule, most abundant in fruit, in the equatorial or hot regions of the globe; they diminish in number, exuberance, and value as we approach the Pole. It should be noted, however, with thankful feelings, that the most widely diffused of all plants are the cereals, such as wheat, barley, and rice, which are also the most useful.

The vegetation which covers the earth, says Buffon, and which is even more closely attached to it than the animals that browse on its surface, are far more interested than they in the question of climate. Not only each country, but each changing degree of temperature, has its particular plants. At the foot of the Alps we find the plants of France and Italy; at their summit, those of the frozen north; and these same northern plants we also find on the crests of the African mountains. On the southern slopes of the lofty range which separates the ancient empire of the Moguls from the kingdom of Cashmere, we find many of the vegetable species of India; on the northern flank, many of those of Europe. It is also from the extremes of climate that we draw our drugs, perfumes, and poisons, and all those plants whose properties are in excess. Temperate climates, on the contrary, produce only temperate things; the mildest of herbs, the wholesomest of vegetables, and the most refreshing of plants and fruits are, like the gentlest of animals, and the most intellectual races of men, the product of the most moderate climates.

How many species of plants does the reader suppose are distributed over the terrestrial surface?

No less than 120,000 have been recognized, named, and described; but there is good reason to believe that thrice this amazing number have yet to be collected and distinguished by our botanists.

You will not expect me to refer to a hundredth part of this wonderful army of trees, shrubs, vegetables, and flowers. All I can do is to point out some of the most important, and to show you where you may reasonably expect to find them. When you are older, you will find that geographical botany is one of the most interesting and delightful of studies.

Well, then, let us turn to *Europe*. This great continent, for botanical purposes, may be divided into three regions—the northern, the central, and the southern.

In the northern, which comprehends Lapland, Iceland, Norway, part of Sweden, and Russia, the landscapes are bare and dreary, and in many places the only vegetation consists of a few mosses or lichens, on which the reindeer feeds. The trees which push their way furthest north are the so-called conifers, such as the fir and pine, and with these the slopes of the northern mountains are abundantly clothed. Quitting this desolate scene, we come to a less frigid atmosphere, where the lime, the ash, and the spreading beech crown the plains with their beauty. In the Highlands of Scotland you will meet with the elegant birch, and up to the 60th degree of latitude ascend the poplar, hazel, and oak; while up to the 70th, the husbandman cultivates his fields of oats and barley.

Coming to the central or temperate region of Europe—that is, to the British Isles, Germany, Holland, Switzerland, Belgium, and the northern districts of France and Italy—we rejoice in a far more abundant distribution of useful and ornamental plants. Our fields are golden with wheat; our gardens abound in flowers, such as roses and lilies, and pinks and cloves; the banks and hedgerows bloom with hawthorns, violets, pansies, hyacinths, primroses, and the like; and in our woods we rejoice beneath the leafy boughs of the common oak, the chestnut, the beech, the elm, the alder, and the willow. Turn to the orchard, and we find the apple, the pear, and the cherry; while, in the most favoured spots, the vine, the fig, the maize, and the mulberry astonish us by their vigour.

In the southern, or Mediterranean region, vegetation attains a happy development. The air is filled with the perfume of the flowers, which include some of the species most distinguished for elegance of form and beauty of

AN ENGLISH ORCHARD.

colour. Here are found the families of the labiatæ, the cistaceæ, the liliaceæ, the borraginaceæ

—names you do not understand as yet, but which, in after-years, will suggest to you a host of pleasant ideas. The evergreen oak, the balmy myrtle, and the glorious rose-laurel adorn the green hill-sides; groves of orange and olive trees diversify the landscape; the hedges are rendered impenetrable by the prickly pear. The cork tree is also a distinctive feature of this plentiful and picturesque vegetation; and we also meet with the citron, the mastic tree, the cypress, and the pomegranate.

Next we betake ourselves to Asia, where we find prevailing a similar threefold division of northern, central, and southern regions.

The vegetation of the north of Asia does not differ largely from that of Europe, though it has, of course, certain distinctive genera. The forests are chiefly composed of the larch, the pine, the poplar, birch trees, service trees, blackberries, alders, willows; while the undershrubs consist of myrtles and Alpine roses. Yet a considerable district of Northern Asia is far richer in aspect than the corresponding latitudes of Northern Europe. The Ouralian forests "present an alternation of a mixture of sharp-edged, round-leaved plants, and other magnificent trees, an assemblage which is completed by masses of brushwood, formed by wild roses, honeysuckles, and junipers; while the *hesperis*, the blue-petalled *polemonium, cortusa, mathiola,* magnificent primroses, and larkspurs, weave a perfect tapestry of flowers; and the water trefoil, with its snowy blossoms and exquisitely-carved leaves, forms the ornament of the marshes."

In Central Asia the distinctive features of its vegetation are strongly marked.

It is here that we gaze enraptured on the splendid beauty of the magnolias, with their large rich leaves and magnificent flowers; on the evergreen foliage and pure white blossoms of the camelia, which is now a favourite in every English conservatory; and on the ornamental foliage of the glossy aucuba. Peculiar to this region of the world is the tea plant, on whose characteristics I have already enlarged; and another native is the graceful paper mulberry, whose leaves afford the necessary nourishment for the silkworm population of China. Here, for the first time, we meet with the elegant palm; the ebony tree, with its delicious cherry-red berries; the Japan medlar, scarcely less delicious; and the black walnut, whose fruit has many valuable recom-

FLOWER OF THE MAGNOLIA.

mendations. Among better known plants, all of which contribute to our supplies of vegetable food, we may name:—

Rice, wheat, barley, oats; the sago palm,

the Caribbean cabbage; apple, pear, orange, quince, apricot, plum, and peach; yam, water-

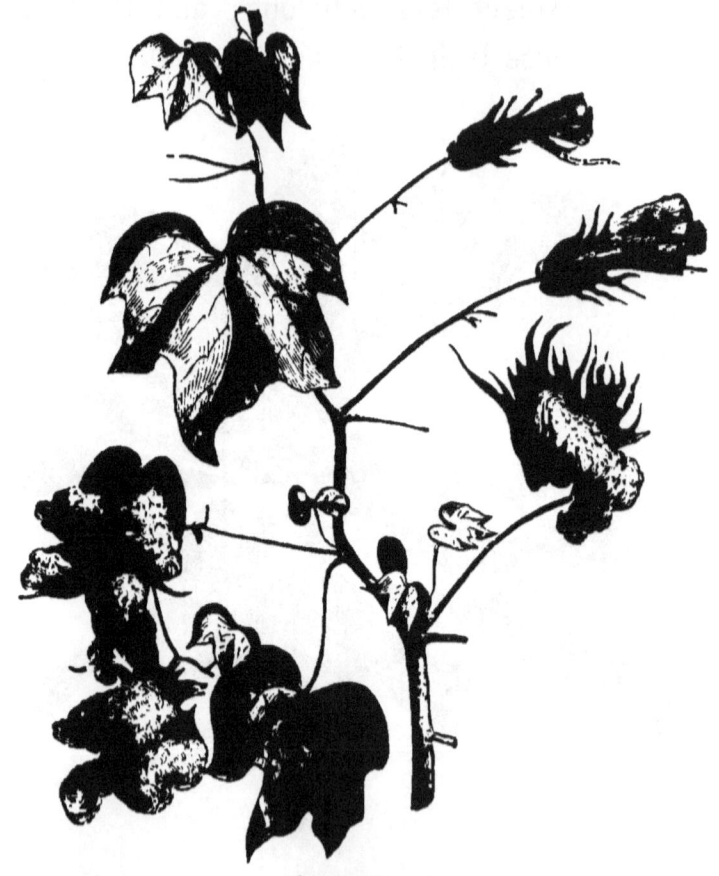

COTTON PLANT.

melon, gourd, cucumber, radish; anise, peas, hemp, and haricot beans.

The all-important cotton plant is one of the most valuable denizens of this favoured region.

"The southern region of Asia," says a recent authority, "comprehends the Indian peninsula. Here *non-tropical* species disappear, or only present themselves very rarely. Tropical groups become more numerous; the trees cease to lose their leaves; ligneous vegetation is much greater than it is without the Tropics; the flowers are larger and more magnificent; climbing, creeping, and parasitic plants increase in number and size. India may be considered the true country of aromatic plants. Nor is the rich soil less fruitful in the production of plants of another order. Trees producing suitable timber for constructive purposes are found there in perfection."

Among the plants which furnish vegetable food we meet with the figs; the sago-palm; the nutmeg tree; the areca; sorghum, or Indian millet; the pungent pepper shrub, so useful as a seasoning; the tamarind; the mango, with its much-extolled and sweet-flavoured fruit; the banana; the luscious guava; the cocoa-nut palm; the mangosteen; and the well-known clove.

But to this region also belong the tall *corypha*

umbraculifera, the gigantic screw-pines, the bamboo, the twining and creeping calamus, the fan-leaved borassus, the cassia, acacia, mimosa, and bignonia.

In Africa three separate regions may very clearly be distinguished; the northern (or Mediterranean and Saharan), the central or tropical, and the southern.

In the Sahara the chief vegetable production is the palm—the date-palm—without which the wandering peoples of the desert would assuredly perish. It is not only important as in itself yielding a most nutritious article of food, but because in its genial shade so many useful plants may be cultivated—as onions, beans, carrots, and cabbages; while the oasis, of which the palm is the centre, and, so to speak, the foundation, also nourishes figs, pomegranates, quinces, pears, apples, and peaches. Pimento is a favourite seasoning with the Arabs. The gombo, the egg plant, and the love-apple are also grown.

Central Africa is imperfectly known; but on the coast flourish immense and almost inaccessible forests of mangroves, extending their roots to the very waters of ocean; of plantains, baobabs, aroids, euphorbias, and aloes. It is

THE MANGROVE.

almost unnecessary to say that the various species of the palm are prominent among the ornaments of African vegetation. The olive-like palm bears a fruit remarkable for the quantity of oil which it contains. Its sap furnishes an excellent wine, and its leaves are much relished by sheep and goats.

The flora of Southern Africa is distinguished by its numerous beautiful varieties of ericas, or heaths; pelargoniums, or geraniums; epacrids, proteas, and ixias. Many of the heaths attain to a height of sixteen feet, and are clothed with blossoms of the most gorgeous character. The stately strelitzias, and the handsome sword-like gladioles, come from the Cape of Good Hope; also the starry-flowered Stapelias, and the mesembryanthemums, or ice plants.

Among the edible plants we may mention the cereals, the sorghums, the tamarind, the banana, the guava, and most of the fruits and vegetables belonging to temperate regions.

Across the broad waters we hasten to America, and first direct our attention to its great northern continent. Here we perceive, as usual, the three regions of northern, central, and southern. The vegetation of the northern calls for no remark, but in the central we meet with many very important varieties. Need we remind even our youngest readers that from North America come the kalmias, rhododendrons, and azaleas which are now so much prized in our English gardens? That in North America flourishes the magnificent tulip tree; and that its groves are planted with maples, limes, and robinias? What shall we say of the American oaks, planes, poplars, elms, ashes, larches, pines, and liquidambars? If we had but the space, how spell-bound we could hold the reader while we dilated on their distinctive merits!

Then turning to the southern region, we would tell him of beautiful yuccas and zamias, of exquisite laurels, of wonderful cactaceæ, of all kinds of fruit-trees, and of the valuable maple, which supplies the well-known maple sugar.

FLOWER OF THE MAPLE.

But leaving him to trace for himself, at a future time, the characters of these innumerable products, we proceed, in our rapid survey, to South America.

Here the great natural divisions are those of

MAPLE-SUGAR BOILING.

the Llanos, the Virgin Forests, the Pampas, and the region of the Andes.

The Llanos are vast grassy plains, chiefly

covered with a dwarf vegetation, but possessing also a few species of palms. Foremost among the latter is the mauritia, which may justly be called the sago tree of America.

In the region of the Andes we meet with the useful cow-tree, which yields a refreshing, wholesome, and nutritious milky juice; and the theobroma cacao, the source of our cocoa and chocolate. Among non-edible plants let us name—we can do no more—the cactus, willow, oak, cinchona, alder, and andromeda. Maize and coffee, the cereals, European vegetables, and the potato are also abundant; and a botanist might study the rich growth of the lower slopes of the Andes for many a year and fail to exhaust its stores.

Perhaps the distinctive feature of the Virgin Forests is their profusion of *lianas*, or creeping and climbing plants, which encircle the trunks and branches of the trees, clasp them round in a stifling embrace, suspend their gay festoons and waving ribbons to the boughs, and weave an inextricable network of glossy leaves and glowing flowers. But the trees are also remarkable from their height, their girth, their unwonted shapes, their evergreen foliage, their numerous varieties of colour.

The vegetation of the Pampas, like that of the Llanos, is chiefly confined to the grasses. In Chili, on the western coast, are immense forests of a peculiar species of palm, the spiny-leaved *Araucaria imbricata*, which grows to the height of a hundred and fifty feet.

In the numerous islands of the Southern Ocean, the two characteristic forms are the cocoa-nut palm and the bread-fruit tree. New Zealand is famous for its gigantic ferns, as well as for the edible taro and sweet potato, and the valuable New Zealand flax. Its forest trees are numerous, and clothed with beautiful foliage.

The flora of Australia differs, in some respects, from that of any other part of the world; but at least one half of it consists of different species of two genera—the eucalyptus or gum-tree, and the acacia. The former is the sacred tree of the natives, and in its shadow they inter their dead.

The sun is now sinking in the west, and I

must conclude my gossip; which, having begun in the corn-fields, has traversed the principal countries of the world to terminate in our own green meadows and shady woods. Wherever we have been, we have found an abundance of Vegetable Food, to gladden the heart and renew the strength of man. We have seen him nowhere confined to a single article—of which he might quickly weary—but furnished with a wonderful variety to stimulate his industry and gratify his tastes. We have seen, too, that each country has its appropriate growth, so that man shall never be left without food in the world. Yet, how little I have told you! How much more remains to be told! If I had had the time, or you were old enough to study the subject more deeply, what countless proofs might I not bring forward, not only of the Divine power and Divine wisdom, but of the Divine love—inexhaustible in its resources, unending in its operation, and boundless in its compassion!—PRAISE BE TO THEE, O FATHER! Thou preparest us food, when Thou hast provided for it!

INDEX.

AFRICA, vegetation of, 151–153.
Almond, the, 121.
America, flora of, 153–157.
Apple, the, 38.
Arrack, 21.
Arrowroot, 70.
Artichoke, the, 62.
Asia, vegetation of, 146–151.
Australia, flora of, 157.

BANYAN, the, 101–103.
Baobab, the, 103–109.
Barley, 13.
Beans, 58.
Brazil nuts, 121.
Bread-fruit tree, the, 98–100.
Broccoli, 64.

CABBAGE, 61–64.
Capers, 71.
Cauliflower, the, 64.
Cherry, the, 35–37.
Chestnut, the, 113–117.
Chocolate, 81, 82.
Cocoa, 81.

Cocoa-nut, the, 83–91.
Coffee, 78–80.
Corn, 16, 17.
Currants, 49.

DATE PALM, the, 91–98.
Dulse, 124.

EUROPE, trees and plants of, 143–146.

FIG-TREE, the, 101.
Flour, 15.
Fungi, edible, 134–137.

GOOSEBERRY, the, 48, 49.
Grapes, 51, 52.
Grass, 12, 13.
Grits, 16.

HAZEL, the, 117, 118.
Hop, the, 122–124.

LEMON, the, 42, 45.
Lichens, 126, 127.

MAIZE, 18, 22.
Millet, 18, 22, 23.
Molasses, 34.
Moss, Icelandic, 128-130.
Muscatels, 51.

NUTMEG, 67.

OATS, 15.
Olives, 68-72.
Onion, the, 58, 59.
Orange, the, 41, 42.

PEACH, the, 39, 40.
Pear, the, 38.
Peas, green, 57, 58.
Pepper, 65-67.
Pine-apple, the, 54.
Plums, 40, 41.
Pomegranate, the, 53.
Potato, the, 55-57.

QUASS, 15.
Quince, 45.

RADISHES, 61
Raisin, the, 50, 51.
Raspberry, the, 47, 43.
Rice, 17-22.
Rock Tripe, 130-133.
Rye, 17-22.

SAFFRON, 70.
Salad, 60.
Sea-kale, 64.
Sherry wine, 68.
Spinach, 62.
Strawberry, the, 45, 46.
Sugar, 24-34.

TAPIOCA, 70.
Tea, 72-77.
Tripe de Roche, 130-133.
Truffle, the, 138.

VINE, the, 49, 50.

WALNUT, the, 118-121.
Wheat, 13.

www.ingramcontent.com/pod-product-compliance
Lightning Source LLC
Chambersburg PA
CBHW030314170426
43202CB00009B/997